Lecture Notes in Mathematics

A collection of informal reports and seminars
Edited by A. Dold, Heidelberg and B. Eckmann, Zürich

Series: Mathematical Systems Theory · Advisers: R. E. K

90

N. P. Bhatia · O. Hajek
Case Western Reserve University, Cleveland, Ohio

Local Semi-Dynamical Systems

1969

Springer-Verlag Berlin · Heidelberg · New York

This work was supported by the National Science Foundation under Grants no. NSF-GP-7447 and NSF-GP-8961. The authors are indebted to Mrs. Elizabeth Roach for her meticulous typing, and to Mr. Charles Allen for the preparation of the diagrams.

Library of Congress Catalog Card Number 70 - 79551 Printed in Germany. Title No. 3696

CONTENTS

0. Introduction 2

1. Local semi-dynamical systems: basic definitions and properties 12

2. Solutions: negative continuation 27

3. Invariance 35

4. Compactness conditions 42

5. Limit sets 49

6. The positive prolongation 59

7. Stability and orbital stability 74

8. Attraction 79

9. Flow near an invariant set 94

10. Liapunov functions 100

11. The start point set 120

12. Minimality; characteristic 0 132

13. Functional-differential equations 141

References 154

Index 156

0. INTRODUCTION

0.1 This paper is devoted to the basic theory of the so-called local semi-dynamical systems. These are objects related to the classical dynamical systems (see 0.2). The differences are that development into positive time only is specified, so that indeterminate behavior into the past is allowed; and it is not assumed that solutions are defined or extendable over all positive times (see 0.6 and 0.7). Such objects arise from a "dynamical" interpretation of functional-differential equations with time-lag, and also of evolution-type partial differential equations (see 0.7 to 0.10). This is in contrast to dynamical systems which mainly arise from ordinary autonomous differential equations, as described below.

0.2 A dynamical system (sometimes called a global bilateral dynamical system) on a topological space X is a continuous mapping $\pi\colon X \times R \to X$ which satisfies the initial value axiom and the group axiom,

$$x\pi 0 = x,$$

$$(x\pi t)\pi s = x\pi(t+s),$$

for all $x \in X$ and $t, s \in R$. (R denotes the real line; the value of π at (x,t) is denoted in operator fashion, by $x\pi t$ instead of $\pi(x,t)$.) For a detailed study of dynamical systems see [15], [6], [10].

0.3 One of the principal motivations for the study of dynamical systems is that they describe completely a certain large class of differential equations.

Consider the differential equations in R^n,

0.3.1 $$\frac{dx}{dt} = f(x,t)$$

with the following assumptions:

0.3.2 The equation is autonomous, i.e. $f(x,t)$ is independent of t; furthermore, $f\colon R^n \to R^n$ is continuous.

0.3.3 The equation has uniqueness of solutions to the initial value problem.

0.3.4 The equation has global existence of solutions: each solution can be extended to a solution defined on the entire real line R.

0.4 Under the conditions just described, with 0.3.1 we may associate a dynamical system π on $X = R^n$ as follows. For any $x \in R^n$, $t \in R$ consider the unique solution $\sigma\colon R \to R^n$ of 0.3.1 with initial value x, i.e. with $\sigma(0) = x$; then define π at (x,t) by setting

0.4.1 $$x\pi t = \sigma(t).$$

From 0.3.3, π is a mapping. From 0.3.4 its domain is $R^n \times R$. The initial value axiom is satisfied trivially. The group axiom follows easily from uniqueness and this characteristic property of autonomous equations: If σ is a solution then so is every t-translate σ_t of σ, defined by

$$\sigma_t(s) = \sigma(s+t).$$

Finally, continuity of π is essentially the classical theorem concerning continuous dependence on initial data of solutions of 0.3.1.

0.5 In the transition from 0.3.1 to π no information is lost: given π we can reconstruct f and 0.3.1. Indeed, observe from 0.4.1 that, for fixed x, $x\pi t$ defines a solution of 0.3.1; in particular,

$$f(x) = f(\sigma(0)) = \frac{d}{dt}\,\sigma(t)\Big|_{t=0} = \frac{d}{dt}\,x\pi t\Big|_{t=0}.$$

For a simple example, consider the linear case,

$$\frac{dx}{dt} = Ax \qquad (A \in R^{n^2}).$$

Then $x\pi t = e^{At}x$.

0.6 The local dynamical systems differ from the global systems just described essentially by relaxing the assumption that π be defined on $X \times R$ to the following: π is a partial map from $X \times R$ into X; its domain is an open subset D of $X \times R$ of the form

$$D = \bigcup_{x \in X} \{x\} \times (\alpha_x, \omega_x)$$

with $-\infty \leq \alpha_x < 0 < \omega_x \leq +\infty$. (For a treatment of these objects see [10].)

For a simple example, the autonomous equation in R,

$$\frac{dx}{dt} = 1 + x^2$$

defines a local (but non-global) dynamical system π:

$$x\pi t = \tan(t + \arctan x);$$

e.g. $\omega_x = \frac{1}{2}\pi - \arctan x$.

0.7 The (global) semi-dynamical systems differ from the global (bilateral) dynamical systems essentially by replacing R by the set R^+ of non-negative reals. Thus development of the system is specified into positive time only.

Finally, the local semi-dynamical systems are obtained by combining the two generalizations just described (for elementary properties of these see [10]). Typical examples of situations where this type of object appears naturally are functional-differential equations (in particular, difference-differential equations) with time lag, and partial differential equations of evolution type. For an example of the first type see 13.3.6.

0.8 Consider the simplest heat-diffusion equation

0.8.1 $$\frac{\partial u}{\partial t} = a^2 \frac{\partial^2 u}{\partial \xi^2}$$

with boundary conditions $u(\pm 1, t) = 0$ for all $t \geq 0$. The corresponding (global) semi-dynamical system π is defined on the Banach space $C_0[-1,1]$ of all continuous functions $x\colon [-1,1] \to R$ with zero end values, $x(\pm 1) = 0$. Then, for any $x \in C_0[-1,1]$ and any $t \geq 0$ consider the solution

$$u\colon [-1,1] \times R^+ \to R$$

of 0.8.1 with initial value x: $u(\xi,0) = x(\xi)$ for $\xi \in [-1,1]$; finally, define $x\pi t \in C_0[-1,1]$ by letting

$$(x\pi t)(\xi) = u(\xi,t) \quad \text{for} \quad \xi \in [-1,1].$$

0.9 Merely for the purpose of obtaining an example where the axioms are easily verified, we pass from the situation of 0.8 to the analogous problem in the Hilbert space $L_2[0,1]$, or rather in the isomorphic sequence space ℓ_2. The corresponding global semi-dynamical system π is then defined as follows: for any $x = (\xi_n) \in \ell_2$ (so that $\Sigma\xi_n^2 < +\infty$) and any $t \in R^+$, let $x\pi t = (\eta_n)$ where

0.9.1 $$\eta_n = \xi_n \cdot \exp - \left(n + \frac{1}{2}\right)^2 \pi^2 a^2 t.$$

Since the second factor is at most 1, we have

0.9.2 $$\|x\pi t\| \leq \|x\|,$$

so that $x\pi t \in \ell_2$ again. The initial value axiom and the semi-group axiom (i.e., the group axiom from 0.2 but required only in R^+) are readily verified. Continuity of π is only slightly more involved. We observe first that $x\pi t$ is linear in x, so that

0.9.3 $$\|x\pi t - y\pi t\| = \|(x-y)\pi t\| \leq \|x-y\|;$$

second, since the second factor in 0.9.1 is actually between 0 and 1, it follows that (for $t \geq s$ in R^+)

0.9.4 $$\|x\pi t - x\pi s\| = \|(x\pi s)\pi(t-s) - x\pi s\| \leq \|x\pi(t-s) - x\|.$$

Now let $x_i \to x$ in ℓ_2, $t_i \to t$ in R^+; then, using 0.9.3 and 0.9.4, we have

$$\|x_1\pi t_1 - x\pi t\| \le \|x_1\pi t_1 - x\pi t_1\| + \|x\pi t_1 - x\pi t\|$$
$$\le \|x_1 - x\| + \|x\pi|t_1 - t| - x\|.$$

Thus to show that $x_1\pi t_1 \to x\pi t$ we merely need to prove that, for fixed $x \in \ell_2$ and $s_1 \to 0$ in R^+, $x\pi s_1 \to x$. This is verified readily from

$$\|x - x\pi s_1\|^2 = \Sigma\xi_n^2\left(1-\exp-\left(n+\tfrac{1}{2}\right)^2 \pi^2 a^2 t\right)$$
$$\le \sum_{n\le k} \xi_n^2 \cdot \left(1-\exp-\left(k+\tfrac{1}{2}\right)^2 \pi^2 a^2 t\right) + \sum_{n>k} \xi_n^2,$$

on choosing appropriate k.

0.10 In contrast to the situation of a bilateral dynamical system, for semi-dynamical systems one may have points x such that $x = y\pi t$ holds for some y only if $t = 0$; such points x are called start points (see 1.6.1 and Section 11). Actually, for functional-differential equations with time lag, most points are start points (see Example 11.6.4). Similarly in the case described in 0.8: if $x = y\pi t$ with $t > 0$, then x is differentiable in $(-1,1)$. Again this is easier verified in the example of 0.9. We observe that if $x = y\pi t$, $t > 0$, then not only $x = (\xi_n) \in \ell_2$, but even $(n\xi_n) \in \ell_2$; thus every $x = (\xi_n) \in \ell_2$ with $\Sigma n^2\xi_n^2 = +\infty$ is a start point.

A second phenomenom may also appear, namely absence

of negative uniqueness (i.e., existence of singular points, see 1.6.2): in general, $x\pi t = y\pi t$ need not imply $x = y$. This is best seen on the functional-differential equation

$$\frac{d}{dt} x(t) = \max\{|x(s)|: t - 1 \leq s \leq t\}.$$

Indeed, take any x, y in $C[-1,0]$ whatsoever, subject only to

$$\|x\| = \|y\| = 1, \quad x(s) = y(s) \quad \text{for} \quad -\tfrac{1}{2} \leq s \leq 0,$$

$$x(0) = y(0) = 1.$$

Then $x\pi t = y\pi t$ for $t \geq \frac{1}{2}$.

0.11 In this paper the authors have attempted two tasks. The first is to carry over (as far as possible at this stage of development) the results of dynamical system theory to the local semi-dynamical systems. In some cases this is merely direct verification; thus e.g., the Weak Attractor Theorem essentially involves only arguments concerning positive time. Elsewhere it is necessary to make minor or rather obvious changes, e.g. in the classification of positive trajectories. Finally, in other cases, the analysis is not at all immediate, and considerable effort was necessary to discover the needed concepts; e.g. the results concerning the behavior of semi-dynamical systems in the neighborhood of compact invariant sets.

The second task is opposed to the first: to examine those aspects of local semi-dynamical system theory in which this markedly differs from that of classical dynamical systems. In particular, this involves absence of global existence, start points and singular points (actually very few results are presented concerning the latter).

0.12 The study of these local semi-dynamical systems, in full generality, was probably initiated in [7] and [8], and systematically continued in a seminar at Case Western Reserve University [9], [4], [11], and at the 1967 Varenna Summer School [5].

0.13 NOTATION. It is more or less standard. R denotes the set of real numbers (with the natural topology and algebraic structures); R^+ and R^- denote the non-negative and non-positive reals, respectively. Euclidean n-space is denoted by R^n (occasionally we use S^n and E^n for the n-sphere and n-ball, respectively). The empty set is denoted by Φ.

The notation in connection with nets is rather lax; x_i may denote a net and also its i-th term; $x_i \to x$ states that the net x_i has a limit x. One further notational convention is described in 1.1.

0.14 A rough idea of the scope of this paper may be had from the list of section headings. The principal mutual dependence of the sections is indicated in the following diagram

$$
\begin{array}{lcccccccccccc}
 & & & & & & 3 & & 5 & & 7 & & 9 \\
0 \rightarrow 13 & \qquad & 1 & \rightarrow & 2 & \rightarrow & & \rightarrow & & \rightarrow & & \rightarrow & \\
 & & & & & & 4 & & 6 & & 8 & & 10 \\
 & & & & & & \downarrow & & & & \downarrow & & \\
 & & & & & & 11 & & & & 12 & &
\end{array}
$$

The index lists the definitions of the main concepts and notational symbols.

1. LOCAL SEMI-DYNAMICAL SYSTEMS:
Basic Definitions and Properties

1.1 DEFINITION. The pair (X,π) is a continuous local semi-dynamical system iff X is a topological space, π is a partial map from a subset of $X \times R^+$ to X (the value of π at (x,t) will be denoted by $x\pi t$; the condition $(x,t) \in$ domain π will be indicated by the clause "$x\pi t$ is defined"), and the following axioms are satisfied:

1.1.1 For every $x \in X$, there is an ω_x, $0 < \omega_x \leqq +\infty$, such that $x\pi t$ is defined iff $0 \leqq t < \omega_x$ (the local existence axiom; ω_x is called the escape time of x).

1.1.2 $x\pi 0 = x$ for each $x \in X$ (the initial value axiom).

1.1.3 $(x\pi t)\pi s = x\pi(t+s)$ whenever the left side is defined (the semi-group axiom).

1.1.4 π: domain $\pi \to X$ is continuous (continuity axiom).

1.1.5 domain π is open in $X \times R^+$ (Kamke's axiom).

In this situation we will also say that π is an lsd-system on (a topological space) X. Occasionally we may say that π acts on X, and that X is the phase space of π.

1.2 Remark.

1.2.1 If domain $\pi = X \times R^+$, then conditions 1.1.1 and 1.1.5 are satisfied trivially and 1.1.3 has a more elegant formulation. In this case π is termed a global semi-dynamical system or in short a gsd-system.

1.2.2 The local existence axiom 1.1.1 may be formulated at greater length as follows. Let $x\pi t$ be defined. Then $x\pi s$ is also defined for all $s \in [0, t+\epsilon]$ for a sufficiently small $\epsilon > 0$. A stronger version of the latter reappears in 1.1.5.

1.2.3 A considerably stronger version of the semi-group axiom is given in Proposition 1.9.

1.2.4 The topology of X enters only into the last two axioms. Their conjunction is equivalent to the following condition: for any set G open in X, the inverse image $\pi^{-1}(G)$ is open in $X \times R^+$. Another formulation is that if $(x_i, t_i) \to (x, t)$ in $X \times R^+$ and $x\pi t$ is defined, then $x_i\pi t_i$ is ultimately defined and $x_i\pi t_i \to x\pi t$.

1.2.5 An alternative form of Kamke's axiom 1.1.5, essentially due to T. Ura [16] is given in Lemma 1.8.

1.3 In the absence of over-riding remarks, there is assumed given an lsd-system π on a topological space X. We introduce several specializations of an lsd-system.

1.3.1 π is termed global iff domain $\pi = X \times R^+$. Clearly π is global iff $\omega_x = +\infty$ for every $x \in X$.

1.3.2 π is said to have negative globality iff for each $x \in X$ and $t \in R^+$, there exists a $y \in X$ with $x = y\pi t$.

1.3.3 π is said to have negative local existence iff for every $x \in X$ and sufficiently small $t > 0$ there exists a $y \in X$ with $x = y\pi t$.

1.3.4 π is said to have negative unicity iff for any $t \in R^+$ and $x_1, x_2 \in X$, $x_1\pi t = x_2\pi t$ holds only if $x_1 = x_2$.

1.3.5 π has the openness property iff for every open subset $G \subset X$ and every $t \in R^+$, the set $G\pi t$ is open.

1.4 In 1.3.5 we have anticipated the formal introduction of some further notation. For subsets $M \subset X$ and $T \subset R^+$ let

1.4.1 $M\pi T = \{x\pi t\colon x \in X,\ t \in T$ and $x\pi t$ is defined$\}$.

If M or T is a singleton, we prefer to write e.g. $x\pi T$ or $M\pi t$ instead of $\{x\}\pi T$ or $M\pi\{t\}$. (Care should be taken in case both M and T are singletons, since then $x\pi t$ may be undefined in the sense of the convention in Definition 1.1, but perfectly well defined, namely as the empty set, in accordance with the present convention.)

1.5 We will need the following easily proved rules of precedence between π and the set-theoretic operations.

1.5.1 $$\left(\bigcup_i M_i\right)\pi\left(\bigcup_j T_j\right) = \bigcup_{i,j} M_i\pi T_j,$$

1.5.2 $$\left(\bigcap_i M_i\right)\pi t \subset \bigcap_i M_i\pi t,$$

1.5.3 $$(M-N)\pi t \supset M\pi t - N\pi t,$$

1.6 DEFINITION. A point $x \in X$ is said to be

1.6.1 a start point iff $x \neq y\pi t$ for any $y \in X$ and $t > 0$.

1.6.2 a singular point iff there exist $x_1 \neq x_2$ and $t > 0$ with $x_1\pi t = x_2\pi t = x$, but $x_1\pi\theta \neq x_2\pi\theta$ for $0 \leq \theta < t$.

As an immediate consequence of these definitions we have

1.7 PROPOSITION. An lsd-systen π on X has

1.7.1 negative local existence iff there are no start points in X.

1.7.2 negative unicity iff there are no singular points in X.

We now give a basic property of the map from X into $R^+ \cup \{+\infty\}$ defined by the assignment $x \to \omega_x$.

1.8 LEMMA. The assignment $x \to \omega_x$ defines a lower semi-continuous map from X into $R^+ \cup \{+\infty\}$, i.e.

$$\omega_x \leqq \liminf_{y \to x} \omega_y.$$

Proof. Take any t with $0 \leqq t < \omega_x$ (from 1.1.1 this is possible, as $\omega_x > 0$). Thus $x\pi t$ is defined, so that according to 1.1.5, $y\pi t$ is defined for y near x. Using 1.1.1 again, we see that $t < \omega_y$ for these y, and hence

$$t \leqq \liminf_{y \to x} \omega_y.$$

Now take $t \to \omega_x$ from the left to obtain the desired result. Q.E.D.

1.9 PROPOSITION. If for any $x \in X$ and $t \in R^+$, $x\pi t$ is defined, then

1.9.1 $$\omega_{x\pi t} = \omega_x - t.$$

Moreover, in this case, the formula 1.1.3 of the semigroup axiom holds iff either side is defined.

Proof.

1.9.2 We first show that $\omega_{x\pi t} \leqq \omega_x - t$. Indeed take any s with $0 \leqq s < \omega_{x\pi t}$. Then by 1.1.1, $(x\pi t)\pi s$ is defined, and then by 1.1.3, $x\pi(t+s)$ is defined. Then again by 1.1.1, $t + s < \omega_x$, i.e., $s < \omega_x - t$. Now keeping x, t fixed, merely take $s \to \omega_{x\pi t}$ with $0 \leqq s < \omega_{x\pi t}$ to obtain the desired inequality.

1.9.3 We now prove $\omega_x - t \leqq \omega_{x\pi t}$. Assume the contrary $\omega_{x\pi t} < \omega_x - t$. Then necessarily $\omega_{x\pi t} < +\infty$. Take any net $t_i \to \omega_{x\pi t}$ with $0 \leqq t_i < \omega_{x\pi t}$. Then by 1.1.1, $(x\pi t)\pi t_i$ is defined, and then by 1.1.3, $(x\pi t)\pi t_i = x\pi(t+t_i)$ is defined. Since $t + t_i \to t + \omega_{x\pi t} < \omega_x$ by assumption, $x\pi(t+\omega_{x\pi t})$ is defined, we have $(x\pi t)\pi t_i = x\pi(t+t_i) \to x\pi(t+\omega_{x\pi t})$. Setting $y = x\pi(t+\omega_{x\pi t})$ and $x_i = (x\pi t)\pi t_i$ we get on using 1.9.2 and Lemma 1.8 the impossible inequality

$$0 < \omega_y \leqq \liminf_{x_i \to y} \omega_{x_i} \leqq \liminf_i \omega_{x\pi t} - t_i = \omega_{x\pi t} - \omega_{x\pi t} = 0.$$

Q.E.D.

The proof of the above Proposition contains the following important result under 1.9.3.

1.10 LEMMA. If $\omega_x < +\infty$ for any $x \in X$, then for any net $t_i \to \omega_x$ with $0 \leqq t_i < \omega_x$, the corresponding net $x\pi t_i$ has no cluster points.

As an immediate application of the above Lemma we have

1.11 THEOREM. Every lsd-system on a compact phase space X is global.

We now introduce the positive trajectory and the right solution through a point x and study their properties.

1.12 DEFINITION. For any $x \in X$,

1.12.1 the set $x\pi R^+$ is called the positive trajectory through (or of) x, and also denoted by C_x or $C(x)$; thus

$$C_x = x\pi R^+ = \{x\pi t\colon 0 \leqq t < \omega_x\}.$$

If $\omega_x = +\infty$ the positive trajectory will be called principal.

1.12.2 the right solution through x is the map ${}_x\pi$,

$${}_x\pi\colon [0,\omega_x) \to X,$$

defined by the assignment ${}_x\pi(t) = x\pi t$.

Evidently ${}_x\pi$ is continuous (see 1.1.4). Further, the positive trajectory through x is the range of the right solution through x,

$$C_x = \text{range}({}_x\pi).$$

Finally, Lemma 1.10 may be stated thus: if $\omega_x < +\infty$, then the cluster set of ${}_x\pi$ at ω_x is empty. An analogous situation will be seen to obtain for the so-called maximal left solutions to be defined in Section 2.

1.13 To study the structure of an individual positive trajectory C_x, the first step is a description of what may be termed the self-intersections of ${}_x\pi$. The resulting classification in terms of the so-called group of periods will then be related with properties of solutions: 1.15; and finally with purely topological properties of trajectories: 1.17. The general situation is slightly simpler if the phase spaces are Hausdorff, and we confine ourselves to this case.

1.14 LEMMA. Let π be an lsd-system in a Hausdorff space X, and let $x \in X$. If $\omega_x < +\infty$ then ${}_x\pi$ is one-to-one. If $\omega_x = +\infty$ then there exists a closed additive subgroup G of R and a $\lambda \in R^+$ such that

$$x\pi t_1 = x\pi t_2, \quad t_1 > t_2$$

if and only if

$$t_1 - t_2 \in G, \quad t_2 \geqq \lambda.$$

G is determined uniquely by these conditions.

Proof.

1.14.1 $x\pi t_1 = x\pi t_2$ implies (see 1.9.1) $\omega_x - t_1 = \omega_x - t_2$, so that $t_1 \neq t_2$ yields $\omega_x = +\infty$ directly. Thus indeed $_x\pi$ is one-to-one if $\omega_x < +\infty$ (one may then take $G = \{0\}$, and $\lambda = \omega_x$). Henceforth we assume that $\omega_x = +\infty$.

1.14.2 For each $t \in R^+$ define

$$A_t = \{s \in R^+ : x\pi t = x\pi(t+s)\};$$

we shall show that A_t is the trace on R^+ of a closed additive subgroup G_t of R. Indeed it will suffice to show that A_t is closed under subtraction in R^+, and that it is a closed subset. For convenience write $y = x\pi t$, so that $s \in A_t$ iff $y = y\pi s$. Let $s_1 \geqq s_2$ in A_t; then

$$y = y\pi s_1 = y\pi(s_2+(s_1-s_2)) = (y\pi s_2)\pi(s_1-s_2) = y\pi(s_1-s_2).$$

Thus indeed $s_1 - s_2 \in A_t$. To verify that A_t is closed, let $s_i \in A_t$ and $s_i \to s$ in R^+. Then

$$y = y\pi s_i \to y\pi s$$

so that $y = y\pi s$ and $s \in A_t$, from the Hausdorff assumption. It is useful to observe that $s \in G_t$ iff $|s| \in A_t$.

1.14.3 We now prove that $t \leqq \theta$ in R^+ implies $G_t \subset G_\theta$. Indeed, let $s \in G_t$; then

$$x\pi t = x\pi(t+|s|)$$

so that

$$x\pi\theta = x\pi(t+\theta-t) = (x\pi t)\pi(\theta-t) = (x\pi(t+|s|))\pi(\theta-t)$$
$$= x\pi(t+|s|+\theta-t) = x\pi(\theta+|s|);$$

i.e. $s \in G_\theta$ as asserted.

1.14.4 We next prove that if $t \leqq \theta$ and $G_t \neq \{0\}$, then $G_t = G_\theta$. In view of 1.14.3 we need only show that $G_\theta \subset G_t$. Thus, let $\sigma \in G_\theta$; by assumption, there is an $s > 0$ in G_t. Then for sufficiently large positive integer n we have $\theta < t + ns$ and then by 1.14.3

$$\sigma \in G_\theta \subset G_{t+ns}.$$

Therefore

$$x\pi(t+ns) = x\pi(t+ns+|\sigma|) = (x\pi t)\pi(ns+|\sigma|).$$

On the other hand $s \in G_t$ implies $ns \in G_t$, and thus

$$x\pi(t+ns) = x\pi t.$$

This together with the last equalities yields $ns + |\sigma| \in G_t$; and since $ns \in G_t$, we have $|\sigma| \in G_t$. Since $\sigma \in G_\theta$ was arbitrary, we have proved that indeed $G_t = G_\theta$.

1.14.5 The result just established shows that there are two cases. Either all $G_t = \{0\}$, whereupon the solution ${}_x\pi$ is one-to-one (we may then put $G = \{0\}$ and $\lambda = 0$ to satisfy the original assertion). Or there exists a $\lambda \in R^+$ and a closed additive subgroup G of R such that

$$G_t = \begin{cases} \{0\} & \text{for } 0 \leqq t < \lambda \\ G & \text{for } \lambda < t. \end{cases}$$

To complete the proof it now suffices to show that even for $t = \lambda$ we have $G_\lambda = G$; or, according to 1.14.3, merely that $G_\lambda \supset G$.

1.14.6 To prove this, take any $s \in G$, so that $s \in G_t$ for all $t > \lambda$. Then

$$x\pi t = x\pi(t+|s|) \quad \text{for} \quad t > \lambda;$$

on taking limits as $t \to \lambda+$ we obtain that $x\pi\lambda = x\pi(\lambda+|s|)$, i.e. that $s \in G_\lambda$. Since $s \in G$ was arbitrary, indeed $G \subset G_\lambda$. This concludes the proof of Lemma 1.14.

The above lemma leads to the following classification and definition.

1.15 DEFINITION. Given an lsd-system π in a Hausdorff space X, and a point $x \in X$, there obtains precisely one of the alternatives (the notation is that of Lemma 1.14):

1.15.1 $G = \{0\}$; C_x is called a non-self-intersecting trajectory.

1.15.2 $\lambda = 0$, G is infinite cyclic; C_x is called a cycle, and also periodic with primitive period (the least positive element of G).

1.15.3 $\lambda = 0$, $G = R$; x and C_x are called (periodic) critical.

1.15.4 $\lambda > 0$ (whereupon $G \neq \{0\}$); C_x is said to lead to a cycle or critical point, according as G is infinite cyclic or $G = R$.

In cases 1.15.2 to 1.15.4, C_x is also called a self-intersecting trajectory.

1.16 The classification just given is formulated in terms of coincidence of values of the solution ${}_x\pi$; we merely have to interpret $s \in G_t$ defined in 1.14 as $x\pi t = x\pi(t+|s|)$, rather as ${}_x\pi(t) = {}_x\pi(t+|s|)$. An equivalent classification, entirely in terms of topological properties of C_x is presented in the following

1.17 PROPOSITION. Let π be an lsd-system in a Hausdorff space X, and $x \in X$. Then

1.17.1 C_x is non-self-intersecting iff it is non-compact.

1.17.2 If C_x is self-intersecting then it is homeomorphic to

1.17.3 $\qquad S^1, \quad \{x\}, \quad$ figure-of-six, $\quad E^1$

according as C_x is periodic with primitive period, or critical, or leads to a cycle or critical trajectory.

Proof.

1.17.4 First let C_x be self-intersecting. By Lemma 1.14,

$$\text{domain }{}_x\pi = [0,\omega_x) = R^+.$$

Consider the canonic factorization of the mapping ${}_x\pi$ (continuous onto)

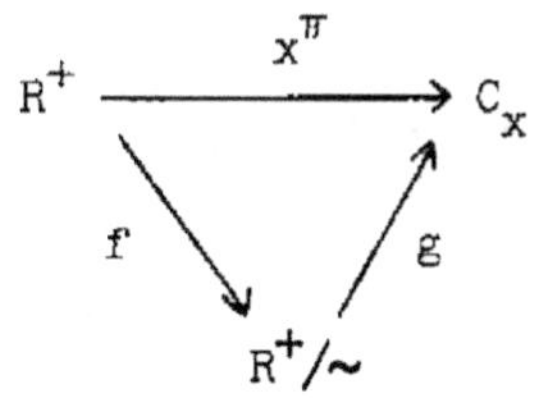

Here the equivalence relation $\sim$ on R^+ is defined by letting $t \sim s$ iff $x\pi t = x\pi s$, and f is the corresponding surjection onto the quotient space. Since ${}_x\pi$ is onto, g is uniquely defined by requiring commutativity, i.e. $g \circ f = {}_x\pi$. All the maps involved are continuous onto, and g is one-to-one. The main result of Lemma 1.14 yields that $R^+/\sim$ is indeed one of the spaces 1.17.3; in particular, it is compact, so that g is a (indeed, the required) homeomorphism. The proof is then completed by observing that the spaces 1.17.3 are mutually non-homeomorphic.

1.17.5 By what we have just proved, a self-intersecting trajectory is compact. Thus it remains to show that the following assumption leads to a contradiction: C_x compact, ${}_x\pi$ one-to-one. From 1.14 again $\omega_x = +\infty$. From compactness, some net $t_i \to +\infty$ has $x\pi t_i$

convergent, necessarily to a point in C_x. Thus $x\pi t_i \to x\pi t$, $t \in R^+$. For like reason there is a convergent subnet of $x\pi(t_i-t-1)$ which is ultimately defined:

$$x\pi(t_j-t-1) \to x\pi s, \quad s \in R^+.$$

Then

$$x\pi t \leftarrow x\pi t_j = (x\pi(t_j-t-1))\pi(t+1) \to (x\pi s)\pi(t+1)$$
$$= x\pi(s+t+1),$$

and hence $t = s+t+1$ since X is Hausdorff and ${}_x\pi$ is one-to-one. This contradiction with $s \in R^+$ concludes the proof of Proposition 1.17.

1.18 Remarks. If $\omega_x < +\infty$ then C_x is non-self-intersecting (see Lemma 1.14) and hence non-compact. It is rather simple to show that in this case C_x is actually homeomorphic to R^+, and even ${}_x\pi$: $[0,\omega_x) \approx C_x$. For principal non-self-intersecting trajectories the following can be proved: $C_x \approx R^+$ iff ${}_x\pi$: $R^+ \approx C_x$, and also iff $x \notin L_x$ (see Section 5).

2. SOLUTIONS: Negative Continuation

In 1.12.2 we defined the right solution $_x\pi$; however, a slightly more general concept will be needed.

2.1 DEFINITION. Given an lsd-system π in a topological space X, an object σ is a solution of π iff the two following conditions hold:

2.1.1 $\sigma\colon I \to X$ with I an interval (i.e., a non-void convex subset) in R^1.

2.1.2 $\sigma(t+s) = \sigma(t)\pi s$ whenever $t \leqq t + s$ in I.

2.2 Remark. The latter condition, in conjunction with 1.9, implies that

2.2.1 $\omega_{\sigma(t)} + t = \omega_{\sigma(s)} + s$ for t, s in I.

The continuity of π yields directly only right-continuity of σ; however, we have the

2.3 PROPOSITION. Every solution is continuous.

Proof. Let $\sigma\colon I \to X$ be a solution of π, let $t_i \to t$ in I. Now, either t is not the left end-point of I, so that $s < t$ for some $s \in I$, and ultimately

2.3.1 $$s \leq t_i = s + (t_i - s);$$

or t is the left end-point, and 2.3.1 again holds on taking $s = t$. In either case, we have, using 2.1.2 twice, that

$$\sigma(t_i) = \sigma(s+(t_i-s)) = \sigma(s)\pi(t_i-s) \to \sigma(s)\pi(t-s) = \sigma(t).$$

Thus $\sigma(t_i) \to \sigma(t)$ and σ is continuous.

2.4 LEMMA. Let $\{\sigma_i | i \in I\}$ be a monotone family of solutions (i.e., for any i, j in I, $\sigma_i \supset \sigma_j$ or $\sigma_j \subset \sigma_i$). Then $\cup\{\sigma_i | i \in I\}$ is a solution.

Proof. Let $\sigma_i\colon I_i \to X$, and set $\sigma = \cup\{\sigma_i | i \in I\}$. Then

$$\text{domain } \sigma = \cup\{\text{domain } \sigma_i | i \in I\} = \cup\{I_i | i \in I\} \ (= J, \ \text{say})$$

is an interval in R'. For any $t \leq t + s$ in J there is an $i \in I$ with t, $t+s$ in I_i, whereupon

$$\sigma(t+s) = \sigma_i(t+s) = \sigma_i(t)\pi s = \sigma(t)\pi s.$$

Hence σ satisfies both the requirements for a solution.

2.5 DEFINITION. Given an lsd-system π in X, and a point $x \in X$, a solution σ of π is

2.5.1 a left solution (through x) iff max domain $\sigma = 0$ (and $\sigma(0) = x$).

2.5.2 a right solution (through x) iff min domain $\sigma = 0$ (and $\sigma(0) = x$).

2.5.3 a left maximal solution iff it is a left solution and it is maximal relative to the property of being a left solution.

2.5.4 a right maximal solution iff it is a right solution and maximal relative to the property of being a right solution.

2.6 Remark. It is easily seen that a right maximal solution through x is precisely the right solution ${}_x\pi$ defined in 1.12.2.

For the purposes of discussing the invariance of subsets of X it is convenient to introduce the notion of a negative trajectory. This we do now.

2.7 DEFINITION. A subset of X is called a negative trajectory (through x) iff it is the range of a left maximal solution (through x). A left solution with domain $\sigma = (-\infty, 0]$ is necessarily left maximal. Such left solutions, and the corresponding negative trajectories, will be called principal.

From 2.4 and the Zorn lemma it follows that, for any monotone system $\{\sigma_i | i \in I\}$ of left solutions there exists a left maximal solution $\sigma \supset \cup \sigma_i$. In particular, for any $x \in X$, there exists a left maximal solution through x (take the monotone system as consisting of a single left solution, defined on the degenerate interval $\{0\}$).

We have now the following "negative" counterpart to Lemma 1.10.

2.8 LEMMA. Let $\sigma: (\alpha,0] \to X$, with $\alpha > -\infty$, be a left solution of π in a Hausdorff space X. Then precisely one of the following alternatives obtains:

2.8.1 There are no cluster points of $\sigma(t)$ as $t \to \alpha+$, whereupon σ is left maximal.

2.8.2 $\lim_{t \to \alpha+} \sigma(t)$ exists; furthermore, the continuous extension of σ to α is a solution of π, so that σ is not left maximal.

Proof.

2.8.3 We shall prove the following. If y is the limit of some net $\sigma(t_i)$ with $t_i \to \alpha+$, then the map $\bar{\sigma}: [\alpha,0] \to X$ defined by

$$\bar{\sigma}(t) = \begin{cases} \sigma(t) & \text{for } \alpha < t \leq 0 \\ y & \text{for } \alpha = t \end{cases}$$

is a solution of π. Indeed, then $\bar{\sigma}$ will be continuous according to Proposition 2.3, so that

$$\lim_{t \to \alpha+} \sigma(t) = \lim_{t \to \alpha+} \bar{\sigma}(t) = \bar{\sigma}(\alpha) = y.$$

2.8.4 Evidently $\bar{\sigma}$ does have an interval for its domain, and it satisfies condition 2.1.2 for most values of the arguments involved. Actually we need only check that

2.8.5 $$\sigma(\alpha+t) = y\pi t \quad \text{for} \quad 0 < t \leqq -\alpha.$$

2.8.6 First consider only $t < \omega_y$. Then

$$\sigma(\alpha+t) \leftarrow \sigma(t_1+t) = \sigma(t_1)\pi t \rightarrow y\pi t$$

and 2.8.5 follows from the Hausdorff condition on X. Having this, we merely have to show that $-\alpha < \omega_y$ (so that $0 < t \leqq -\alpha$ in 2.8.5 will imply $t < \omega_y$). Again take $0 < t < \omega_y$, so that 2.8.5 holds. Then from 1.9.1 and 2.2.1

$$\omega_y = \omega_{\sigma(\alpha+t)} + t = \omega_{\sigma(0)} - \alpha > -\alpha$$

since always $\omega_x > 0$. This completes the proof.

The following proposition now provides a classification of left-maximal solutions. The routine proof is left to the reader.

2.9 PROPOSITION. Let π be an lsd-system in a Hausdorff space X; let σ be a left-maximal solution, and $N = \text{range}\ \sigma$ the corresponding negative trajectory. Then precisely one of the following alternatives obtains:

2.9.1 σ and N are principal, i.e. domain $\sigma = (-\infty, 0]$.

2.9.2 domain $\sigma = [\alpha, 0]$ with $-\infty < \alpha \leq 0$; then $\sigma(\alpha)$ is a start point. (In this case we will say that σ and N lead from the start point $\sigma(\alpha)$.)

2.9.3 domain $\sigma = (\alpha, 0]$ with $-\infty < \alpha < 0$; then $\sigma(t)$ has no cluster points as $t \to \alpha+$.

2.10 COROLLARY. If $\overline{N}$ is compact (e.g., X compact), then 2.9.1 and 2.9.2 are the only alternatives. If π has negative local existence, then 2.9.1 and 2.9.3 are the only alternatives.

2.11 Remark. The classification of left-maximal solutions naturally provides a classification of the negative trajectories. It should be observed that distinct left-maximal solutions can have coinciding ranges, i.e. define the same negative trajectory. However, this happens if and only if the trajectory contains periodic points; the situation is then perhaps clear from the classification of positive trajectories described in Section 1, and it is unnecessary to go into details here.

We close this section with some definitions which are needed later.

2.12 DEFINITION. If N is a negative trajectory through x, then $N \cup C_x$ is called a complete trajectory through x ($N \cap C_x \neq \{x\}$ is not excluded); it is called principal iff both N and C_x are principal.

2.13 DEFINITION. Let $x \in X$ and σ be a left-maximal solution through x. We write

2.13.1 $$\alpha_x = -\sup\{t: x \in X\pi t\},$$

2.13.2 $$\alpha_\sigma = \inf \text{ domain } \sigma,$$

and call α_x the negative escape time of x and α_σ the negative escape time of σ.

It is clear that

2.13.3 $$-\infty \leqq \alpha_x \leqq \alpha_\sigma \leqq 0.$$

Evidently x is a start point iff $\alpha_x = 0$. Furthermore, π has negative local existence iff $\alpha_x < 0$ for all $x \in X$; π has negative globality iff $\alpha_x = -\infty$ for all $x \in X$. It may well happen that $\alpha_x < \alpha_\sigma$ for all left-maximal solutions σ through x (and this even when π has negative globality). Obviously,

$$\alpha_{x\pi t} \leqq \alpha_x - t$$

whenever $x\pi t$ is defined.

Finally, we introduce the definition of the (negative) funnel through x.

2.14 DEFINITION. Given $x \in X$ and $s \leqq t$ in R^+,

2.14.1 the set $F_x = F(x) = \{y: x \in C_y\}$ is called the (negative) funnel through x.

2.14.2 the set $F_s^t(x) = \{y: x \in y\pi[s,t]\}$ is a section of the funnel through x.

2.14.3 the set $F_t(x) = F_t^t(x) = \{y: x = y\pi t\}$ is a cut of the funnel through x.

Finally, for any subset $M \subset X$ we define

$$C(M) = \cup\{C_x | x \in M\}, \ \ldots, \ F_t(M) = \cup\{F_t(x) | x \in M\}.$$

It is easily seen that for any $x \in X$,

2.14.4 $F_x = \cup\{N | N$ is a negative trajectory through $x\}$.

3. INVARIANCE

In this section we will be treating two types of invariance; the strong and the weak versions. They coincide if the lsd-system has negative unicity. Without further mention we assume given an lsd-system π on a Hausdorff space X.

3.1 DEFINITION. $M \subset X$ is called

3.1.1 positively invariant iff $M = M\pi R^{+}$, or equivalently, iff

$$M\pi t \subset M \quad \text{for all} \quad t \in R^{+}.$$

3.1.2 (strongly) negatively invariant iff $X - M$ is positively invariant.

3.1.3 (strongly) invariant iff it is both positively and negatively invariant.

The following two lemmas have straightforward proofs (for the notation F_x see 2.14.1).

3.2 LEMMA. $M \subset X$ is positively invariant iff $C_x \subset M$ for each $x \in M$. The set M is negatively invariant iff $F_x \subset M$ for each $x \in M$, or equivalently, iff for each $x \in M$, $N \subset M$ for each negative trajectory N through x.

3.3 LEMMA.

3.3.1 Φ and X are invariant.

3.3.2 The union and intersection of positively invariant sets is positively invariant.

3.3.3 To each $M \subset X$ there exists the least positively invariant set containing M (usually called the positively invariant hull of M, viz. $M\pi R^+$), and also the largest positively invariant subset of M (the positively invariant kernel of M).

Similar assertions hold for negative invariance and for invariance.

3.4 LEMMA.

3.4.1 If M is positively invariant then so is $\overline{M}$ and also each path-component of M.

3.4.2 If M is negatively invariant then so is $\text{Int}\, M$ and also each path-component of M.

3.4.3 If M is invariant then its boundary ∂M is positively invariant.

Proof.

Proof of 3.4.1: In any case $\overline{M} \subset \overline{M}\pi 0 \subset \overline{M}\pi R^+$. The opposite inclusion follows from the continuity of the mapping π:

$$\overline{M}\pi R^+ \subset \overline{M\pi R^+} = \overline{M}.$$

Second, $C_x \subset M$ whenever $x \in M$; since C_x is path-connected, it must be within the same path-component of M as x.

Proof of 3.4.2: The first assertion follows from 3.4.1 by taking complements. For the second observe that $F_x \subset M$ whenever $x \in M$. Now F_x is the union of negative trajectories through x and the latter are path-connected and have a common point x. Thus each F_x is path-connected, and therefore F_x and x are in the same path component of M.

Proof of 3.4.3: This follows from $\partial M = \overline{M} \cap \overline{X-M}$, 3.4.1 and 3.3.2.

3.5 Remark. It is well worth noting that if M is positively invariant then $\text{Int } M$ need not be positively invariant. This is in contrast to the situation in dynamical systems. Again in 3.4.3, ∂M need not be invariant or even weakly invariant (see Definition 3.7).

Closed positively invariant sets will be of considerable interest in the stability theory. We give a handy characterization of these.

3.6 LEMMA. A closed set M is positively invariant if and only if, for every $x \in \partial M$, there exists an $\epsilon > 0$ with $x\pi[0,\epsilon) \subset M$.

Proof. If M is positively invariant, then the condition holds, e.g. with $\varepsilon = \omega_x$. For the converse part, assume that M is not positively invariant. Then there exists $x \in M$, $t > 0$ with $x\pi t \notin M$. Since M is closed and ${}_x\pi$ continuous, there exists a last $s \in R^+$ with

$$x\pi[0,s] \subset M, \quad s < t.$$

Then $y = x\pi s \in \partial M$, and the condition cannot hold for this y, by the construction of s. Q.E.D.

We now introduce weak invariance.

3.7 DEFINITION. A subset $M \subset X$ is weakly negatively invariant iff, for each $x \in M$, there is a negative trajectory N through x with $N \subset M$. A subset simultaneously positively and weakly negatively invariant will be termed weakly invariant.

The following lemma has an immediate proof.

3.8 LEMMA.

3.8.1 Negative invariance implies weak negative invariance.

3.8.2 The union of weakly negatively invariant subsets is again such.

3.8.3 If M_1 is negatively invariant and M_2 weakly negatively invariant, then $M_1 \cap M_2$ is weakly negatively invariant.

Our next result is the negative counterpart of 3.6. It will enable us to prove among other matters, the corresponding parallel to Lemma 3.4; however, this will be delayed to Section 4.

3.9 PROPOSITION. A closed set M is weakly negatively invariant iff, for every non-start point $x \in \partial M$, there exists y and $\epsilon > 0$ such that

$$y\pi\epsilon = x, \quad y\pi[0,\epsilon] \subset M.$$

Proof.

3.9.1 Evidently every weakly negatively invariant and closed subset satisfies the condition. Conversely, assume the condition, and take any $x \in M$. Now consider the system of all left solutions σ through x which also satisfy range $\sigma \subset M$. Using 2.4 and Zorn's Lemma, there is a solution σ maximal relative to these requirements; set $N = \text{range } \sigma$. Then $N \subset M$, and it only remains to prove that σ is a left-maximal solution in the sense of 2.5.

3.9.2 If σ is principal, then it is indeed left-maximal. In the remaining cases there is an $\alpha \leqq 0$

such that domain σ has one of the forms

$$[\alpha, 0], \quad (\alpha, 0].$$

3.9.3 In the first case $\sigma(\alpha)$ is a start point, so that σ is left-maximal. Indeed, assume not. Then either $\sigma(\alpha) \in \text{Int } M$, and σ could be continued negatively within $\text{Int } M \subset M$, a contradiction with the maximality required of σ. Or $\sigma(\alpha) \in \partial M$, and then the assumed condition again yields this contradiction.

3.9.4 In the second case σ is again left-maximal. Indeed, in the opposite case we obtain (from 2.3 and closedness of M) that $\lim_{t \to \alpha+} \sigma(t)$ exists and is in M; this with Lemma 2.8 contradicts the maximality of σ. This concludes the proof.

3.10 COROLLARY. The intersection of nested compact weakly invariant sets without start points is weakly invariant.

Proof. Let $\{M_i\}$ be a system of sets as described; we propose to verify the condition from 3.9 for $M = \bigcap M_i$. Since the phase space is Hausdorff, the compact set M is closed. Take any $x \in \partial M \subset M \subset M_i$; by assumption, for each i there is a negative trajectory $N_i \subset M_i$ through x. Since M_i is compact without start points, N_i is principal (see 2.10). Thus, for any fixed $\varepsilon > 0$ and each i, there exist $y_i \in N_i$ with

$$y_i\pi\varepsilon = x, \quad y_i\pi[0,\varepsilon] \subset N_i \subset M_i.$$

Now take a convergent subnet $y_j \to y$ (as the ordering relation on the index set take the inclusion for the M_i); then

$$y\pi\varepsilon = x, \quad y\pi[0,\varepsilon] \subset \bigcap M_j = \bigcap M_i = M.$$

This completes the proof.

4. COMPACTNESS CONDITIONS

In this section we collect those results which explicitly require certain compactness conditions on the phase space X and or subsets of X. These will be needed in later sections.

We recall that a topological space X is called rim-compact (or semi-compact) iff each point has a base of neighborhoods with compact boundaries; in general this is strictly weaker than local compactness (but, e.g., in normed linear spaces the two requirements are equivalent). It is easily established that every Hausdorff rim-compact space is regular. Besides this we need the following easily established result.

4.1 LEMMA. Let X be a Hausdorff space and $M \subset X$ be compact. If a net x_i, $x_i \in X$, is ultimately in every neighborhood of M, then every subnet x_j of this net has a cluster point in M.

We assume as before that an lsd-system π on a Hausdorff space X is given. As a simple but useful consequence of Lemma 1.10 we have

4.2 LEMMA. If $M \subset X$ is compact and $C_x \subset M$, then C_x is principal.

As an obvious consequence of 4.1 and 4.2 we have

4.3 LEMMA. Let $M \subset X$ be compact. Let for an $x \in X$, the net $x\pi t$, with t in the naturally directed set $[0, \omega_x)$, be ultimately in every neighborhood of M. Then $C_x \cup M$ is compact; consequently, C_x is principal.

We now establish a less obvious result.

4.4 PROPOSITION. Let U be a neighborhood of $M \subset X$, with both M and ∂U compact. Then there exists a neighborhood V of M and a $t > 0$ such that

$$F_0^t(V) \subset U.$$

Proof. Assume the contrary; then it is easily shown that there exist

$$x_i \notin U, \quad t_i \to 0, \quad x_i\pi t_i \to x \in M.$$

Since U is a neighborhood of M, ultimately

$$x_i\pi t_i \in U \not\ni x_i;$$

thus the connected sets $x_i\pi[0,t_i]$ must intersect ∂U, and we have

$$x_i\pi s_i \in \partial U, \quad 0 \leqq s_i < t_i.$$

We conclude, first, that $s_i \to 0$; and second, since ∂U is compact, that there is a convergent subnet $x_j\pi s_j \to y \in \partial U$. However, then

$$x \leftarrow x_j\pi t_j = (x_j\pi s_j)\pi(t_j-s_j) \to y\pi 0 = y$$

and $x = y$ (X being Hausdorff). This contradicts the fact that $x \in M \subset \text{Int } U$, and $y \in \partial U$.

4.5 LEMMA. Suppose that $(x_i,t_i) \to (x,t)$ in $X \times R^+$, and that all $x_i\pi[0,t_i]$ are in a compact subset M. Then all $x_i\pi t_i$ and $x\pi t$ are defined and $x_i\pi t_i \to x\pi t$; moreover, $x\pi[0,t] \subset M$.

Proof.

4.5.1 Indeed if some $x_i\pi t_i$ is not defined, then $t_i \geqq \omega_{x_i} < +\infty$ and therefore $C_{x_i} = x_i\pi[0,\omega_{x_i})$ $= x_i\pi[0,t_i]$ is in the compact set M. This contradicts Lemma 4.2. Thus each $x_i\pi t_i$ is defined.

4.5.2 If $\omega_x > t$, then indeed $x\pi t$ is defined and $x\pi[0,t] \subset M$ follows from continuity of π and closedness of M directly. The proof will therefore be completed by showing that $\omega_x > t$. Assume the contrary. Then for every s, $0 \leqq s < \omega_x$ ($\leqq t$), $t_i > s$ ultimately. Therefore $x_i \pi s$ is defined ultimately and is in M; moreover, $x_i \pi s \to x\pi s$ by continuity of π (note that $x\pi s$ is defined by assumption on s). Thus $x\pi s \in M$ as M is closed, and since $s \in [0, \omega_x)$ was arbitrary, $C_x \subset M$. This contradicts Lemma 4.2, since $\omega_x < +\infty$ by assumption and M is compact. Q.E.D.

We now state without proof the following useful lemma. A proof can easily be given with the help of the above result.

4.6 LEMMA. Let $M \subset X$ be compact and $x \in M$. The set M contains a principal negative trajectory through x if and only if for each $\epsilon > 0$ there is a $y \in M$ with

$$y\pi\epsilon = x, \quad \text{and} \quad y\pi[0, \epsilon] \subset M.$$

As a final result we have an application of 4.5.

4.7 THEOREM. For every x in a locally compact phase space X, and sufficiently small $t > 0$, the sets $F_0^t(x)$ and $F_t(x)$ are compact.

Proof. Take a compact neighborhood U of x; then according to 4.4

$$F_t(x) \subset F_0^t(x) \subset U$$

for small $t > 0$. Hence and from Lemma 4.5, both the sets considered are closed; since U is compact, these sets are also such. Q.E.D.

4.8 Remarks. We observe that in any case F_0^t is pathwise connected; any $y \in F_0^t$ can be joined to x by the path $y\pi[0,t_y]$, where $y\pi t_y = x$. However, in contrast with the situation of differential equations, the cuts $F_t(x)$ are often not connected. Of course, if x is not singular (see 1.6.2) then the cuts $F_t(x)$ are singletons (or void if x is a start point) for small $t > 0$, and hence connected. Equally obviously, the cuts need not be connected if the phase space X is not rather well behaved, e.g. if it is the triad as in figure 4.8.1 (as the time parameter take the real coordinate).

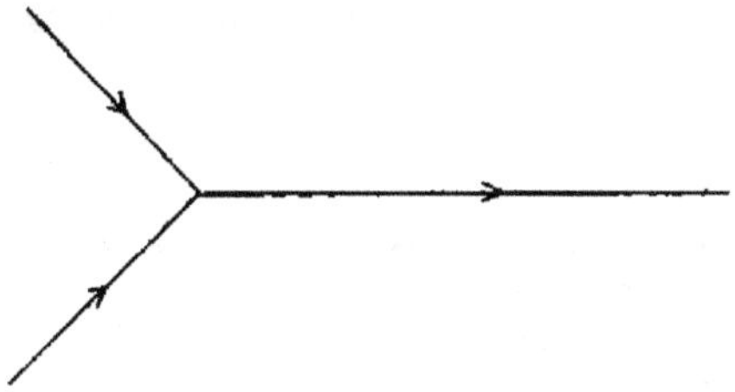

Figure 4.8.1. Triad Phase Space.

We will close this section with a theorem on weak negative invariance which corresponds to the result 3.4.1 for positive invariance.

4.9 THEOREM. Let X be locally compact and $M \subset X$ be a weakly negatively invariant set containing no start points. Then $\overline{M}$ is weakly negatively invariant and contains no start points.

Proof. Let $y \in \overline{M}$ and let U be a compact neighborhood of y. Then there is a net y_i in $\text{Int } U \cap M$ with $y_i \to y$. By negative weak invariance of M, there are left maximal solutions σ_i, $\sigma_i(0) = y_i$, with $N_i = \text{range } \sigma_i \subset M$. Now, there are two possibilities. Either there is a subnet σ_j with $N_j \subset U$, or there is a subnet σ_j with $N_j \cap \partial U \neq \Phi$. In the first case the σ_j are principal (U is compact and M contains no start points). Thus there are points $x_j \in U$ with $x_j\pi[0,1] \subset U \cap M$ and $x_j\pi 1 = y_j$.

We may assume $x_j \to x \in U \cap \overline{M}$, so that by Lemma 4.5 $x\pi 1$ is defined, and

$$y \leftarrow y_j = x_j\pi 1 \to x\pi 1, \quad y = x\pi 1.$$

Moreover, $x\pi[0,1] \subset \overline{M}$. In the second case there are points $x_j \in \partial U$ and $t_j > 0$ with $x_j\pi[0,t_j] \subset U$ and $x_j\pi t_j = y_j$. Now either $t_j \leqq 1$ ultimately, or there is a subnet $x_k\pi(t_k-1)$ in U. In the latter case the situation is similar to that of the first case, and so there is a point $x \in \overline{M}$ with $x\pi[0,1] \subset \overline{M}$ and $x\pi 1 = y$. In the former case there is a subnet $x_k \to x \in \partial U$, $t_k \to t$ with $0 \leqq t \leqq 1$, so that

$$y \leftarrow y_k = x_k\pi t_k \to x\pi t$$

yields $y = x\pi t$; and from $x_i\pi[0,t_i] \subset U \cap M$ it follows that $x\pi[0,t] \subset \overline{M}$. Since $y \in \text{Int } U \setminus x$ we must have $t > 0$. Thus we have shown that no point of $\overline{M}$ is a start point and the conditions of Proposition 3.9 hold. Thus $\overline{M}$ is weakly negatively invariant. Q.E.D.

4.10 Remark. This theorem does not hold in rim-compact spaces X, even for global semi-dynamical systems as may be seen on simple examples.

5. LIMIT SETS

The principal aim of this section is to study the properties of the set $\overline{C}_x$ which will be denoted by K_x or $K(x)$. It will be shown that under rather general conditions the set K_x contains a largest weakly invariant set which is related to the positive limit set L_x to be presently defined. We assume given an lsd-system π on a Hausdorff space.

5.1 DEFINITION. For any $x \in X$, the set

5.1.1 $\{y: x\pi t_i \to y$ for some net $t_i \to +\infty\}$

is called the positive limit set of x and denoted by L_x or $L(x)$.

5.2 Remark.

5.2.1 If for an $x \in X$, $\omega_x < +\infty$, then the net $x\pi t_i$, $t_i \to +\infty$, is not defined ultimately. Hence for such x, $L_x = \Phi$. Equivalently, if $L_x \neq \Phi$, then C_x is principal.

5.2.2 We will have occasion to consider sets $C_{x\pi t}$, $K_{x\pi t}$ for a given $x \in X$ and $t \in R^+$. If indeed $x\pi t$

is defined, then the sets $C_{x\pi t}$ and $K_{x\pi t}$ are perfectly defined. In case $x\pi t$ is not defined then we set $C_{x\pi t} = K_{x\pi t} = \Phi$.

An alternate description of the set L_x is given by the following easily proved proposition.

5.3 PROPOSITION. For an $x \in X$,

$$L_x = \cap\{K_y | y \in C_x\} = \cap\{K_{x\pi n} | n \text{ a positive integer}\}.$$

The following theorem follows easily from the definition.

5.4 THEOREM. For any $x \in X$,

5.4.1 $K_x = C_x \cup L_x$,

5.4.2 K_x and L_x are closed and positively invariant,

5.4.3 $L_x = L_y$ for $y \in F_x \cup C_x$.

Proof. 5.4.1 is an easy consequence of the definitions and the Hausdorffness of the phase space X. 5.4.2 follows from the fact that C_x, and consequently K_x, (by using 3.4.1) is positively invariant for any $x \in X$; and then using 3.3.2 to get the result for L_x. Lastly for 5.4.3 we observe that if $y \in F_x \cup C_x$, then either $y \in C_x$ or $x \in C_y$. It is therefore sufficient to show that if $y \in C_x$ then $L_x = L_y$. But this is almost self-evident from 5.3. Q.E.D.

In contrast to the situation in the theory of dynamical systems [15], [6], [10] where the positive limit sets are invariant, it can easily be shown on examples that in the case of an lsd-system the positive limit set of a point need not even be weakly invariant and this even if negative unicity obtains. The next theorems will therefore be devoted to obtain conditions under which this does happen.

5.5 THEOREM. Let K_x be compact for an $x \in X$. Then L_x is a non-void compact connected weakly invariant set. Moreover, L_x contains no start points and every negative trajectory in L_x is principal.

Proof. Since K_x is compact, $C_x \subset K_x$ is principal (Lemma 4.2). Thus $K_{x\pi n}$ is a compact connected non-void set for each positive integer n. Since $K_{x\pi n} \supset K_{x\pi m}$ for $m \geqq n$, we conclude from $L_x = \cap\{K_{x\pi n} | n = 1,2,\ldots\}$ and elementary topological considerations [13, F, p. 163] that L_x is compact connected and non-void. That it is positively invariant has already been established (5.4.2), moreover since L_x

is compact every positive trajectory in L_x is principal (see 4.2). To prove that L_x is negatively weakly invariant, contains no start points, and every negative trajectory in L_x is principal we will use Lemma 4.6. So let $y \in L_x$. Then there is a net $x\pi t_i \to y$, $t_i \to +\infty$. Choose any $\varepsilon > 0$ and consider the net $y_i = x\pi(t_i-\varepsilon)$ which is clearly ultimately defined (C_x is principal) and in the compact set K_x. There is therefore a convergent subnet $y_j \to z$. Indeed, since $y_j = x\pi(t_j-\varepsilon)$ and $t_j - \varepsilon \to +\infty$, $z \in L_x$. On the other hand $y_i\pi[0,\varepsilon] \subset K_x$. From Lemma 4.5,

$$y \leftarrow x\pi t_j = y_j\pi\varepsilon \to z\pi\varepsilon ,$$

and therefore $z\pi\varepsilon = y$, and say by positive invariance of L_x, $z\pi[0,\varepsilon] \subset L_x$. Since $\varepsilon > 0$ was arbitrary, the conditions of Lemma 4.6 are satisfied and L_x contains a principal trajectory through every point $y \in L_x$. Q.E.D.

Another condition implying weak invariance of L_x is given below.

5.6 THEOREM. If X is locally compact, then for any $x \in X$ the set L_x is weakly invariant and contains no start points.

Proof. For $L_x = \Phi$ the result is trivial. For $L_x \neq \Phi$ we need only establish that any $y \in L_x$ is not a start point and L_x is weakly negatively invariant. Take any $y \in L_x$. There is a net $x\pi t_i \to y$ with $t_i \to +\infty$. Consider the net $x\pi(t_i-1)$ which is ultimately defined and $t_i - 1 \to +\infty$. Take a compact neighborhood U of y. Then either $(x\pi(t_i-1))\pi[0,1] \subset U$ ultimately, so that we may assume $x\pi(t_i-1) \to z \in L_x$; and then

$$y \leftarrow x\pi t_i = (x\pi(t_i-1))\pi 1 \to z\pi 1$$

and $z\pi 1 = y$. Or there is a subnet $x\pi(t_j-1)$ such that the curves $(x\pi(t_j-1))\pi[0,1]$ intersect ∂U for each j. Thus there exists s_j, $t_j - 1 \leqq s_j \leqq t_j$, with $x\pi s_j \in \partial U$ and $s_j \to +\infty$, and $x\pi[s_j,t_j] \subset U$. Since ∂U is compact we may assume that the net $x\pi s_j \to z$. Then clearly $z \in L_x$. Moreover, since $0 \leqq t_j - s_j \leqq 1$ we may assume that $t_j - s_j \to \epsilon \geqq 0$.

Then by Lemma 4.5 $z\pi\varepsilon$ is defined, $z\pi[0,\varepsilon] \subset U$, and $y = z\pi\varepsilon$. Since $y \neq z$ $(y \in \text{Int } U, \ z \in \partial U)$ we must have $\varepsilon > 0$. Thus $y \in L_x$ is not a start point, and moreover L_x satisfies conditions of Proposition 3.9. Thus L_x is weakly invariant and contains no start points. Q.E.D.

A slightly different result is the following

5.7 THEOREM. Let X be rim-compact, and L_x be compact. Then L_x has all the properties asserted in Theorem 5.5.

The above theorem may be proved using Theorem 5.5 together with the easily established following lemma.

5.8 LEMMA. Let X be rim-compact. Then for any $x \in X$, L_x is compact non-void if any only if K_x is compact.

Another important property of the set L_x is in the

5.9 THEOREM. Let K_x be compact (or let X be rim-compact and L_x compact non-void). Then for every neighborhood U of L_x there is a $T > 0$ such that $x\pi t \in U$ for $t \geqq T$.

This property is often denoted by

$$x\pi t \to L_x \quad \text{as} \quad t \to +\infty.$$

We leave the simple proof to the reader.

We now establish some properties of L_x when L_x is not compact. For this we need the following result from topology [12, p. 37].

5.10 LEMMA. Let M be a Hausdorff continuum, i.e. non-void compact connected. Let $U \subset M$ be open non-void. Then every component of U has a limit point in $\overline{U} - U$.

5.11 THEOREM. Let X be locally compact. Let for an $x \in X$, L_x be not connected. Then L_x has no compact component.

Proof. Let $\tilde{X} = X \cup \{\infty\}$ be a one-point compactification of X. Let for any $z \in X$, $K_z^* = \overline{C}_z$ where the closure is taken in $\tilde{X}$. Define $L_z^* = \cap\left\{K_y^* \middle| y \in C_z\right\}$. Since K_z^* is compact connected non-void one easily concludes that L_z^* is compact connected and non-void. However, note that $K_x^* = K_x \cup \{\infty\}$, since K_x is not compact; consequently, $L_x^* = L_x \cup \{\infty\}$. Thus L_x is an open subset of the Hausdorff continuum L_x^*. By virtue of Lemma 4.8 every component of L_x has a limit point in $L_x^* - L_x = \{\infty\}$ and is therefore not compact.

Q.E.D.

It is possible to characterize L_x in locally compact spaces and we give the following result.

5.12 THEOREM. Let X be locally compact. Then for any $x \in X$, L_x is the largest weakly invariant subset of K_x which contains no start points.

The proof depends on Theorem 4.9 and is left to the reader.

5.13 Remark. It would appear that Theorem 5.6 and Theorem 5.12 may hold in rim-compact spaces. This is in general not the case. However, if the lsd-system π is global then Theorems 5.6 and 5.12 hold even in rim-compact spaces. In this connection we give the following example.

5.14 Example. Consider the rim-compact space obtained by deleting the closed segments

$$P_n = \left\{(x,y) \in R^2: x = 1, -\frac{1}{n+1} \geqq y \geqq -\frac{1}{n}\right\},$$

for all odd integers $n > 0$, from the euclidean plane R^2. On this space one can easily define an lsd-system π whose trajectories are shown in figure 5.14.1 (e.g.

this is obtained by restricting the phase space of the dynamical system in example 3.10, [15, p. 343], Nemytskii-Stepanov to the set $R^2 - \cup\{P_n | n$ odd positive integer$\}$. For any point $(x,y) \neq (0,0)$, $-1 < x < 1$, the positive limit set is the set $\{(x,y): |x| = 1\} - \cup\{P_n | n$ positive odd integer$\}$. However, this contains the start point $(1,0)$. The system is not global.

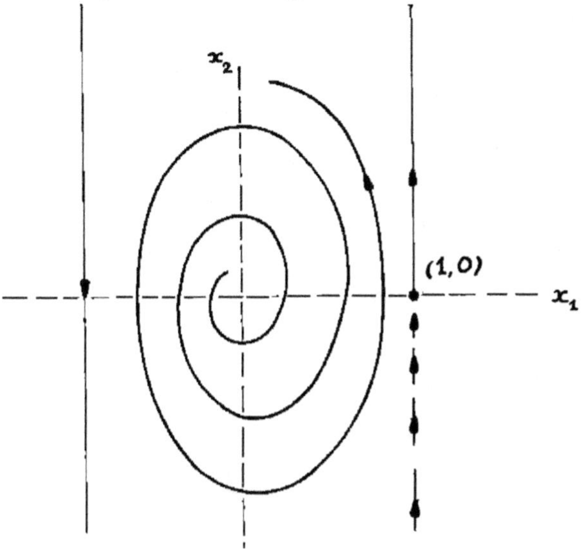

Figure 5.14.1

5.15 Remark. By considering left-maximal solutions through a point x one may define a negative limit set of such a solution σ to be the set

$$\{y: \sigma(t_1) \to y \text{ for some } t_1 \to -\infty\}$$

and denoted by L_σ^-. It is clear that $L_\sigma^- = \Phi$ if σ is not principal. For these sets it is possible to establish all the properties of the positive limit sets expressed earlier in this section. Of course if N is the corresponding negative trajectory, then $\overline{N} = N \cup L_\sigma^-$ replaces 5.4.1 and 5.4.3 has no meaning.

6. THE POSITIVE PROLONGATION

The positive prolongation of a point $x \in X$ and the prolongational limit set of a point $x \in X$ will be defined in this section and their properties studied. This, together with the limit set theory, will form fundamental tools for the study of stability and attraction properties in the next three sections. An lsd-system π on a Hausdorff space X is assumed given.

6.1 DEFINITION. For any $x \in X$ the set

6.1.1 $\{y\colon x_i \pi t_i \to y \text{ for some nets } x_i \to x,\ t_i \geq 0\}$

is called the positive prolongation of x and denoted by D_x or $D(x)$. The set

6.1.2 $\{y\colon x_i \pi t_i \to y \text{ for some nets } x_i \to x,\ t_i \geq 0 \text{ and } \liminf t_i \geq \omega_x\}$

is called the positive prolongational limit set of x and denoted J_x or $J(x)$.

The following proposition gives alternate descriptions of the sets D_x and J_x. The easy proof is left out.

6.2 PROPOSITION. For any $x \in X$,

6.2.1 $D_x = \cap\{\overline{U\pi R^+} | U$ is a neighborhood of $x\}$.

6.2.2 $J_x = \cap\{D_y | y \in C_x\} = \cap\{D_{x\pi t} | 0 \leqq t < \omega_x\}$.

The following theorem gives simple properties of the sets D_x, J_x. The easy proofs are left out.

6.3 THEOREM. For any $x \in X$,

6.3.1 $D_x = C_x \cup J_x$.

6.3.2 D_x and J_x are closed and positively invariant.

6.3.3 $J_x \supset L_x$.

6.4 Remark. Contrary to the situation for dynamical systems where (the corresponding sets are denoted by D_x^+ and J_x^+) $D_y^+ \supset D_x^+$, $J_y^+ = J_x^+$ holds for $y \in C_x^+$, the sets D_x and J_x do not have these properties as may be seen on easy examples. However, $J_x \subset J_{x\pi t}$ for $t \in [0, \omega_x)$. Further, the set J_x need not be weakly invariant in contrast to the corresponding set J_x^+ being invariant in dynamical systems [6]. The set J_x can be characterized in yet another useful way if the space X is assumed to be rim-compact. We shall need this result in proving invariance properties of J_x in locally compact spaces.

6.5 PROPOSITION. Let X be rim-compact. Then for any $x \in X$ with $\omega_x < +\infty$,

6.5.1 $$J_x = \{y: x_i\pi t_i \to y \text{ for some nets } x_i \to x \text{ and } \liminf t_i > \omega_x\}.$$

Proof. It is sufficient to show that if $x_i \to x$ and $t_i \to \omega_x$, then the net $x_i\pi t_i$ has no cluster points. Assume the contrary, that $x_i\pi t_i \to y$. Since the net $x\pi t$, $t \in [0,\omega_x)$, has no cluster points (Lemma 1.10) there is a t, $0 \leqq t < \omega_x$, such that $y \notin C_{x\pi t}$. (Indeed note that all $C_{x\pi t}$, $t \geqq 0$, are closed sets because of $\omega_x < +\infty$.) Choose a closed neighborhood U of y with compact boundary which does not intersect $C_{x\pi t}$ (this is possible since X is rim-compact, hence also regular). By continuity of π we have $x_i\pi t \to x\pi t \notin U$; moreover, since $(x_i\pi t)\pi(t_i-t) = x_i\pi t_i \to y \in \text{Int } U$, the curves $(x_i\pi t)\pi[0,t_i-t]$ intersect ∂U. Thus there are s_i, $0 \leqq s_i \leqq t_i - t$ with $(x_i\pi t)\pi s_i \in \partial U$. As ∂U is compact, there is a subnet $(x_j\pi t)\pi s_j \to z \in \partial U$ with $s_j \to s$, $0 \leqq s \leqq \omega_x - t = \omega_{x\pi t}$ (note that $t_i - t \to \omega_x - t$). Now there are two possibilities, both leading to a contradiction. If $s < \omega_x - t$, then

$(x_j \pi t)\pi s_j \to (x\pi t)\pi s = z$, so that $z \in C_{x\pi t}$ which contradicts $z \in \partial U$. If $s = \omega_x - t$, then

$$y \leftarrow x_j \pi t_j = ((x_j \pi t)\pi s_j)\pi(t_j - t - s_j) \to z\pi 0 = z$$

as $t_j - t - s_j \to 0$. Thus $y = z$ as X is Hausdorff, a contradiction to the fact that $z \in \partial U$ and $y \in \operatorname{Int} U$. Q.E.D.

Another very useful result using a compactness condition is the following.

5.5 LEMMA. Let $U \subset X$ have compact ∂U. Let $x \in U$, $y \notin U$, and $y \in D_x$. Then there is a $u \in \partial U$ such that $u \in D_x$ and $y \in D_u$.

Proof. Indeed if x or y is in ∂U the result is trivial. So let $x \in \operatorname{Int} U$, $y \notin \overline{U}$. Since $y \in D_x$, there is a net $x_i \pi t_i \to y$ with $x_i \to x$ and $t_i \geqq 0$. We may assume that all $x_i \in \operatorname{Int} U$, and all $x_i \pi t_i \notin \overline{U}$. Then each curve $x_i \pi [0, t_i]$ intersects ∂U so that there are s_i, $0 < s_i < t_i$, with $x_i \pi s_i \in \partial U$. As ∂U is compact, there is a subnet $x_j \pi s_j \to u \in \partial U$. Then clearly $u \in D_x$ and since $x_j \pi t_j = (x_j \pi s_j)\pi(t_j - s_j) \to y$ with $t_j - s_j > 0$, we have $y \in D_u$. Q.E.D.

We are now ready to deal with certain properties of sets D_x and J_x which need some kind of compactness conditions.

6.7 THEOREM. Let $x \in X$ and let there exist a neighborhood U of x with $\overline{U\pi R^+}$ compact. Then D_x and J_x are both compact and connected. Moreover, J_x is weakly invariant, contains no start points, and every trajectory in J_x is principal.

Proof.

6.7.1 We first prove that D_x is compact and connected. D_x being a closed subset of the compact set $\overline{U\pi R^+}$, it is itself compact. To verify connectedness assume that there are non-void disjoint compact sets P, Q with $D_x = P \cup Q$. Let U_P and U_Q be any disjoint neighborhoods of P and Q respectively. Let $x \in P$ and choose $y \in Q$. Since $y \in D_x$, there is a net $x_i\pi t_i \to y$ with $x_i \to x$ and $t_i \geqq 0$. We may assume $x_i \in \text{Int}(U_P \cap U)$ and $x_i\pi t_i \in \text{Int } U_Q$. Thus the curves $x_i\pi[0,t_i]$ intersect ∂U_P, so that there are s_i, $0 < s_i < t_i$, with $x_i\pi s_i \in \partial U_P$. Since $\overline{U\pi R^+}$ is positively invariant, we have indeed $x_i\pi s_i \in \partial U_P \cap \overline{U\pi R^+}$

which is compact. There is therefore a convergent subnet $x_j \pi s_j \to u \in \partial U_P \cap \overline{U\pi R^+}$. Clearly $u \in D_x = P \cup Q$ by definition, which contradicts $u \in \partial U_P$. Thus D_x is connected.

6.7.2 We shall now prove that J_x is compact and connected. Compactness follows from J_x being a closed subset of the compact set $\overline{U\pi R^+}$. If J_x is not connected, then there are non-void disjoint compact sets P, Q with $J_x = P \cup Q$. Since K_x is compact, L_x is non-void compact and connected (Theorem 5.5), and therefore $L_x \subset P - Q$ or $L_x \subset Q - P$. Let $L_x \subset P$. Then $D_x = K_x \cup J_x = (K_x \cup P) \cup Q$. We claim that the sets $K_x \cup P$ and Q are non-void disjoint compact. Indeed if $(K_x \cup P) \cap Q \neq \Phi$, then $Q \cap C_x \neq \Phi$ as $K_x \cup P = C_x \cup L_x \cup P = C_x \cup P$ and $P \cap Q = \Phi$. Note however that Q is positively invariant (since J_x is such and Q is the union of components), so that $C_{x\pi t} \subset Q$ for some $t \geqq 0$. Then $L_{x\pi t} \subset Q$. But $L_x = L_{x\pi t}$ for $t \geqq 0$ and $L_x \subset P$. This contradiction shows that J_x is connected.

6.7.3 Since J_x is compact and positively invariant, each positive trajectory C_y with $y \in J_x$ is contained in J_x and hence is principal.

6.7.4 We now show that for each $y \in J_x$ and any $\varepsilon > 0$ there is a $z \in J_x$ with $z\pi\varepsilon = y$. Then $z\pi[0,\varepsilon] \subset J_x$ follows from positive invariance of J_x. This will show by Lemma 4.6 that J_x contains a principal negative trajectory through each $y \in J_x$, and consequently contains no start points. So let $y \in J_x$, $\varepsilon > 0$. There is a net $x_i\pi t_i \to y$ with $x_i \to x$ and $t_i \to +\infty$ (note $\omega_x = +\infty$ as C_x is principal). Consider the net $x_i\pi(t_i-\varepsilon)$ which is ultimately in the compact set $\overline{U\pi R^+}$. There is a convergent subnet $x_j\pi(t_j-\varepsilon) \to z \in J_x$ as $t_j - \varepsilon \to +\infty$ and $x_j \to x$. From

$$y \leftarrow x_j\pi t_j = (x_j\pi(t_j-\varepsilon))\pi\varepsilon \to z\pi\varepsilon$$

it follows that $y = z\pi\varepsilon$. Q.E.D.

6.8 Remark. In the above theorem it is not asserted that D_x has a compact neighborhood. In locally compact spaces X the condition of the theorem is satisfied whenever D_x is compact. Hence

6.9 COROLLARY. If X is locally compact, then the conclusions of Theorem 6.7 hold whenever either D_x is compact or J_x is compact and non-void.

The last statement in the corollary above is evident from

6.10 LEMMA. Let X be rim-compact and $x \in X$. Then D_x is compact if and only if J_x is compact non-void.

Proof. We only prove the non-trivial part that if J_x is compact non-void, then D_x is compact. We first claim that if J_x is compact non-void then L_x is compact non-void. For if $L_x = \Phi$, then $C_x \cap J_x = \Phi$, since otherwise some $x\pi t$ would be in the compact positively invariant set J_x, and hence $L_x = L_{x\pi t} \neq \Phi$; and C_x is closed since $K_x = C_x \cup L_x = C_x$. Since X is rim-compact (hence regular) there is a closed neighborhood U of J_x with compact boundary such that $C_x \cap U = \Phi$. By Lemma 6.5 there is a $u \in \partial U$ with $u \in D_x$. But $u \notin C_x \cup J_x$ by our choice of the neighborhood U: a contradiction to $D_x = C_x \cup J_x$. Hence L_x is compact non-void. Then K_x is compact by Lemma 5.8 and so $D_x = K_x \cup J_x$ is compact.

We would like to observe that the conclusions of Corollary 6.9 hold even in rim-compact spaces; however, a separate proof is required.

6.11 THEOREM. Let X be rim-compact and $x \in X$. Let either D_x be compact or J_x be compact non-void. Then both D_x and J_x are connected; moreover, J_x contains no start points, is weakly invariant, and contains only principal trajectories.

Proof. Using Lemma 6.9 the proof of connectedness can be carried over from Theorem 6.6. Due to compactness of J_x we need only prove that J_x contains no start points and is weakly invariant.

6.11.1 Let $y \in J_x$. There is a net $x_i \pi t_i \to y$ with $x_i \to x$ and $t_i \to +\infty$ (note that Lemma 6.9 shows that K_x is compact, so that C_x is principal, hence $\liminf t_i \geqq \omega_x = +\infty$). Consider the ultimately defined net $x_i \pi(t_i - 1)$. If $z_i = x_i \pi(t_i - 1)$, then

$$y\pi 1 \leftarrow z_i \pi 1 = x_i \pi t_i \to y$$

so that $y = y\pi 1$ (note that $y\pi 1$ is defined as C_y, being a subset of the compact set J_x, is principal). Otherwise there is a closed neighborhood U of y with compact boundary and a subnet $z_j \notin U$ with $z_j \pi 1 = (x_j \pi(t_j - 1))\pi 1 = x_j \pi t_j$ in U. Thus the curves $z_j \pi[0,1]$ intersect ∂U and there are s_j, $0 \leqq s_j \leqq 1$, with $z_j \pi s_j \in \partial U$. We may assume that $s_j \to s$ and

$z_j\pi s_j \to z \in \partial U$. Since $z_j\pi s_j = (x_j\pi(t_j-1))\pi s_j = x_j\pi(t_j-1+s_j)$ and $t_j - 1+s_j \to +\infty$, we have $z \in J_x$. Furthermore,

$$z\pi(1-s) \leftarrow (z_j\pi s_j)\pi(1-s_j) = z_j\pi 1 \to y$$

as $z\pi(1-s)$ is defined ($C_z \subset J_x$ is principal). Consequently $z\pi(1-s) = y$; but $s \neq 1$, otherwise $z = z\pi 0 = y$ contradicts $y \in \text{Int } U \not\ni z$. Thus we have shown that J_x contains no start points, and the condition of Proposition 3.9 is satisfied. Hence J_x is weakly invariant and contains no start points. Q.E.D.

Our next theorem contains observations on the sets D_x and J_x in locally compact spaces without the requirement that D_x be compact.

6.12 THEOREM. Let X be locally compact and $x \in X$.

6.12.1 The set J_x is weakly invariant and contains no start points.

6.12.2 D_x has no compact components if it is disconnected; and analogously for J_x.

Proof.

6.12.3 Proof of 6.12.1. We need only show that J_x is negatively weakly invariant and contains no start points. Let $y \in J_x$. Then there is a net $x_i \pi t_i \to y$ with $x_i \to x$ and $t_i \to \tau$, where either $\tau = +\infty$ or $\tau > \omega_x$ if $\omega_x < +\infty$ (Proposition 6.5). Let U be a compact neighborhood of y and choose $\varepsilon > 0$ either arbitrarily or with $\varepsilon < \tau - \omega_x$ if ω_x is finite. Consider the net $x_i \pi (t_i - \varepsilon)$ which is ultimately defined. Either the curves $(x_i \pi (t_i - \varepsilon)) \pi [0, \varepsilon]$ are frequently in U or ultimately they intersect ∂U. In the first case there is a subnet $x_j \pi (t_j - \varepsilon)$ with the curves $(x_j \pi (t_j - \varepsilon)) \pi [0, \varepsilon]$ in U and $x_j \pi (t_j - \varepsilon) \to z \in U$ (since U is compact). Since $t_j - \varepsilon \to \tau - \varepsilon$ and $\tau - \varepsilon = +\infty$ or $\tau - \varepsilon > \tau - (\tau - \omega_x) = \omega_x$ if ω_x is finite, we have $z \in J_x$, and moreover by Lemma 4.5 and positive invariance of J_x, we have $z\pi[0, \varepsilon] \subset J_x$. Also,

$$y \leftarrow x_j \pi t_j = (x_j \pi (t_j - \varepsilon)) \pi \varepsilon \to z \pi \varepsilon$$

shows that $y = z\pi\varepsilon$. In the second case, the curves $(x_i \pi (t_i - \varepsilon)) \pi [0, \varepsilon]$ intersect ∂U ultimately. Thus we may assume that there exist $s_i \geqq 0$ with $0 \leqq t_i - s_i \leqq \varepsilon$ such that $x_i \pi s_i \in \partial U$, $x_i \pi [s_i, t_i] \subset U$, $x_i \pi s_i \to z \in \partial U$, and $t_i - s_i \to s$ with $0 \leqq s \leqq \varepsilon$. Then

$$y \leftarrow x_i\pi t_i = (x_i\pi s_i)\pi(t_i - s_i) \rightarrow z\pi s,$$

so that $y = z\pi s$, and $z \in J_x$ (since $s_i \rightarrow +\infty$ or $s_i \rightarrow \tau - s > \tau - \epsilon > \omega_x$ according as $t_i \rightarrow +\infty$ or $t_i \rightarrow \tau$, and in either case $z \in J_x$). But $s \neq 0$ for otherwise $y = z\pi 0 = z$ will contradict $y \in \text{Int } U \not\ni z$. Hence $s > 0$. Thus we have proved that for each point $y \in J_x$ there is an $\epsilon > 0$ and $z \in J_x$ with $y = z\pi\epsilon$ and $z\pi[0,\epsilon] \subset J_x$. Therefore J_x contains no start points and by Proposition 3.9 is weakly negatively invariant.

6.12.4 Proof of 6.12.2. Consider the one point compactification $\tilde{X} = X \cup \{\infty\}$ of X. For any $x \in X$ define $D^*(x) = \cap\{\overline{U\pi R^+} | U$ is a neighborhood of $x\}$ and $J^*(x) = \cap\{D^*_y | y \in C_x\}$, where C_x has the same meaning but the closures are taken in the space $\tilde{X}$. If either D_x or J_x is disconnected then neither D_x nor J_x is compact (otherwise they would be connected). It is easy to see that if D_x or J_x is not compact, then $D^*_x = D_x \cup \{\infty\}$ and D^*_x is compact and connected. The same argument as in the proof of Theorem 5.11 using Lemma 5.10 shows that no component of D_x is compact. Similarly for the case of J_x. Q.E.D.

The hypothesis of local compactness in the above results may be weakened to rim-compactness in the case of global semi-dynamical systems. Thus we have

6.13 THEOREM. Let X be rim-compact and π a global semi-dynamical system on X. Then for any $x \in X$ the set J_x is weakly invariant and contains no start points.

The proof of this theorem may be taken over from 6.11.1, where the text within parentheses may be omitted.

We are now ready for the following characterization of J_x.

6.14 PROPOSITION. Let either π be an lsd-system on a locally compact X, or π be a global semi-dynamical system on a rim-compact X. Then for any $x \in X$, J_x is the largest weakly negatively invariant subset of D_x which contains no start points.

As a final useful tool we prove

6.15 LEMMA. If $y \in L_x$, then $J_x \subset J_y$.

This will follow from

6.16 LEMMA. If $\{y, z\} \subset L_x$, then $y \in J_z$. Consequently $D_y = J_y$ for every $y \in L_x$.

Proof. There exist nets $x\pi t_i \to y$, $x\pi\tau_i \to z$ with $t_i \to +\infty$ and $\tau_i \to +\infty$. We may assume, by taking appropriate subnets if necessary, that $t_i - \tau_i \to +\infty$. Then

$$y \leftarrow x\pi t_i = (x\pi\tau_i)\pi(t_i-\tau_i)$$

and clearly $y \in J_z$. Since y, z in L_x were arbitrary, we conclude $y \in J_y$ for each $y \in L_x$. Now $D_y = C_y \cup J_y = J_y$ follows from positive invariance of J_y and $y \in J_y$.

Proof of Lemma 6.15. For $y \in L_x$ and any neighborhood U of y, there is a $t \in R^+$ such that U is a neighborhood of $x\pi t$. Then $\overline{U\pi R^+} \supset D_{x\pi t} \supset J_x$. Thus $J_x \subset \cap\{\overline{U\pi R^+} | U$ is a neighborhood of $y\} = D_y$. Since $D_y = J_y$ for every $y \in L_x$, the result follows.

We recall that $J_x \subset J_{x\pi t}$, so that we have the following formally stronger result.

6.17 COROLLARY. If $y \in K_x$, then $J_x \subset J_y$.

In Section 12 we shall need the following result which is clearly related to 6.15. The simple proof is omitted.

6.18 LEMMA. Let σ be a left maximal solution through $x = \sigma(0)$, and $y \in L_\sigma^-$. Then $K_x \subset J_y$.

As a final result we give

6.19 THEOREM. Let $x_i \to x$, $y_i \to y$. If $y_i \in D(x_i)$, then $y \in D(x)$; if $y_i \in J(x_i)$, then $y \in J(x)$.

Proof.

6.19.1 Assume $y_i \in D(x_i)$. If $y \notin D(x)$, then there is an open neighborhood U of x such that $y \notin \overline{U\pi R^+}$. Then there is a subnet $x_j \in U$ with $y_j \notin \overline{U\pi R^+}$. But as U is open, it is a neighborhood of each x_j and so $D(x_j) \subset \overline{U\pi R^+}$. This contradicts $y_j \notin \overline{U\pi R^+}$. Hence we must have $y \in D(x)$.

6.19.2 Now assume $y_i \in J(x_i)$. Then indeed by the above result, $y \in D(x)$. If $y \notin J(x)$, then there is a t, $0 < t < \omega_x$, such that $y \notin D(x\pi t)$. Consequently, there is an open neighborhood U of $x\pi t$ with $y \notin \overline{U\pi R^+}$. Since $\liminf_{y \to x} \omega_y \geqq \omega_x$, $x_i\pi t$ is ultimately defined and continuity of π implies $x_i\pi t \to x\pi t$. Consequently ultimately $x_i\pi t \in U$ and therefore $D(x_i\pi t) \subset \overline{U\pi R^+}$. Hence $y_i \in J(x_i) \subset D(x_i\pi t) \subset \overline{U\pi R^+}$ holds ultimately. But this contradicts $y_i \to y \notin \overline{U\pi R^+}$. Q.E.D.

7. STABILITY AND ORBITAL STABILITY

In this section we introduce the notion of stability and orbital stability of a set and characterize these in terms of prolongations. As before, there is assumed given an lsd-system π on a Hausdorff space X.

7.1 DEFINITION. A set $M \subset X$ is said to be

7.1.1 stable if and only if for each $x \notin M$, $y \in M$, there exist neighborhoods U of x and V of y with $U \cap (V\pi R^+) = \Phi$.

7.1.2 orbitally stable if and only if it has arbitrarily small positively invariant neighborhoods.

The following lemma expresses a necessary condition for stability and orbital stability.

7.2 LEMMA. Let $M \subset X$ be either stable or orbitally stable. Then M is positively invariant.

Proof.

7.2.1 Let M be stable. Let $x \notin M$. For each $y \in M$, $x \notin y\pi R^+ = C_y$ follows from the definition of stability. Thus $x \notin C(M) = \cup\{C_y | y \in M\}$. Hence $C(M) \subset M$ and M is positively invariant.

7.2.2 Let M be orbitally stable. If $C(M) \not\subset M$ then there is an $x \in C(M) - M$. But then $X - \{x\}$ is a neighborhood of M which contains no positively invariant neighborhood; a contradiction to orbital stability of M. Q.E.D.

In general stability and orbital stability are independent concepts. Our next proposition characterizes stability in terms of prolongations.

7.3 PROPOSITION. A set $M \subset X$ is stable iff $D(M) = M$.

Proof.

7.3.1 Let M be stable. For each $x \notin M$, $y \in M$ there are neighborhoods U of x and V of y with $U \cap V\pi R^+ = \Phi$. Consequently $x \notin \overline{V\pi R^+}$ and hence

$$x \notin \cap\{\overline{V\pi R^+} | V \text{ a neighborhood of } y\} = D_y.$$

Since $y \in M$ was arbitrary, we have $x \notin D(M)$, so that $D(M) \subset M$. This shows that $D(M) = M$.

7.3.2 Let $D(M) = M$. Let $x \notin M$, $y \in M$. Then $x \notin D_y \subset D(M)$. Since $D_y = \cap\{\overline{V\pi R^+} | V$ a neighborhood of $y\}$, we conclude the existence of a neighborhood V of y with $x \notin \overline{V\pi R^+}$. But then $X - \overline{V\pi R^+} = U$ is a neighborhood of x and $U \cap V\pi R^+ = \Phi$. Q.E.D.

In general no simple characterization of orbital stability in terms of prolongations seems possible. However, in some cases orbital stability and stability are equivalent. This is expressed in the next theorem.

7.4 THEOREM. Let X be rim-compact. Let M be a closed subset of X with a compact boundary. Then M is stable iff M is orbitally stable.

Proof.

7.4.1 Let M be stable, so that $D(M) = M$ by 7.3. Now assume that M is not orbitally stable. Choose a closed neighborhood U of M with compact boundary. Then $V\pi R^+ \cap \partial U \neq \Phi$ for every neighborhood V of M. The set of neighborhoods of M is directed by inclusion, and we have proved that for each neighborhood V of M there is a point $x_V \in V$ with $x_V\pi R^+ \cap \partial U \neq \Phi$. Since M is positively invariant we must have $x_V \notin M$. Since ∂M is compact, the net x_V has a cluster point in ∂M. Thus there is a net $x_i \to x \in \partial M$ with $x_i\pi t_i \in \partial U$, $t_i \geqq 0$, and $x_i\pi t_i \to u \in \partial U$ as ∂U is compact. Then $u \in D_x$, a contradiction as

$D_x \subset D(M) \not\ni u$.

7.4.2 Now let M be orbitally stable. Then for each $x \notin M$, there are disjoint open neighborhoods U of M

and V of x (since M is closed with compact boundary and X is Hausdorff). By orbital stability of M, there is a neighborhood W of M with $W\pi R^+ \subset U$. Thus $V \cap W\pi R^+ = \Phi$. Since W is a neighborhood of each $y \in M$, M is stable. Q.E.D.

7.5 Remark. Note that only in 7.4.1 we used the full force of our hypothesis. 7.4.2 shows that orbital stability implies stability in the case of closed sets with compact boundaries in Hausdorff spaces. In regular spaces X one can show by the same argument that, for closed sets M, orbital stability implies stability.

The following restatement of the result in 7.3 and 7.4 is useful.

7.6 THEOREM. Let X be rim-compact and M a closed subset of X with compact boundary. Then M is orbitally stable if and only if $D(M) = M$.

We now turn our attention to stability of an lsd-system π on a given phase space X.

7.7 DEFINITION. An lsd-system π is said to be

7.7.1 stable iff, for every $x \in X$, K_x is stable,

7.7.2 orbitally stable iff, for every $x \in X$, K_x is orbitally stable.

It is clear that an lsd-system π is stable if and only if, for each $x \in X$, $D(K_x) = K_x$. Since in general $D(K_x) \neq D_x$ (equality does not hold even for dynamical systems: see [6]), the following theorem is of some interest.

7.8 THEOREM. An lsd-system π is stable iff $K_x = D_x$ for each $x \in X$.

Proof.

7.8.1 Let π be stable. Then $K_x = D(K_x) \supset D_x$ for any $x \in X$. Consequently $K_x = D_x$ for each $x \in X$.

7.8.2 Let $K_x = D_x$ for each $x \in X$. Then

$$D(K_x) = \cup\{D_y | y \in K_x\} = \cup\{K_y | y \in K_x\} = K_x.$$

Thus K_x is stable. Q.E.D.

7.9 Remark. One may define another kind of stability of π by requiring every C_x to be stable, i.e. $D(C_x) = C_x$. This turns out to be equivalent to $D_x = C_x$ (for dynamical systems this is trivial as there $D^+(C_x^+) = D_x^+$ hold always).

8. ATTRACTION

In this section the following two nets will play an important role.

8.1 For any fixed $x \in X$, the net $x\pi t$ with t in the ordered subset $[0,\omega_x)$ of reals, where ω_x is the escape time of x $(0 < \omega_x \leqq +\infty)$.

8.2 For any $x \in X$ and neighborhood U of x, the net $U\pi t$ with t in the ordered set R^+.

We shall use the following notation in dealing with the nets in 8.1 and 8.2.

8.3 NOTATION. Given any $x \in X$, U a neighborhood of x and $M \subset X$, we say that

8.3.1 the net $x\pi t$ is frequently in M iff for each $t \in [0,\omega_x)$ there is a τ with $t \leqq \tau < \omega_x$ such that $x\pi\tau \in M$.

8.3.2 the net $x\pi t$ is ultimately in M iff there is a $T \in [0,\omega_x)$ with $x\pi[T,\omega_x) \subset M$.

8.3.3 the net $U\pi t$ is ultimately in M (before ω_x) iff there is a T with $0 \leqq T < \omega_x$ such that $U\pi t \subset M$ for $t \geqq T$.

Note that $U\pi t$ may but need not be empty for $t \geqq \omega_x$.

We are now ready for some definitions. An lsd-system π on a Hausdorff X is assumed given.

8.4 DEFINITION. Let $x \in X$ and $M \subset X$. Then the point x is

8.4.1 weakly attracted to M iff the net $x\pi t$ is frequently in every neighborhood of M.

8.4.2 attracted to M iff the net $x\pi t$ is ultimately in every neighborhood of M.

8.4.3 strongly attracted to M iff for every neighborhood U of M there is a neighborhood V of x such that the net $V\pi t$ is ultimately in U.

8.4.4 uniformly attracted to M iff there is a neighborhood U of x such that the net $U\pi t$ is ultimately in every neighborhood of M.

With a given set $M \subset X$ one can associate a region of attraction corresponding to each of the notions of attraction introduced above.

8.5 DEFINITION. Let $M \subset X$. Define

8.5.1 $A_w(M) = \{x \in X: \ x \text{ is weakly attracted to } M\}$.

8.5.2 $A(M) = \{x \in X: \ x \text{ is attracted to } M\}$.

8.5.3 $A_s(M) = \{x \in X: \ x \text{ is strongly attracted to } M\}$.

8.5.4 $A_u(M) = \{x \in X: \ x \text{ is uniformly attracted to } M\}$.

The sets $A_w(M)$, $A(M)$, $A_s(M)$, $A_u(M)$ are called the region of weak attraction, attraction, strong attraction, and uniform attraction, of M, respectively.

The following proposition is self-evident.

8.6 PROPOSITION. For any $M \subset X$,

8.6.1 $A_w(M) \supset A(M) \supset A_s(M) \supset A_u(M)$.

8.6.2 $A_w(M)$ and $A(M)$ are invariant.

8.6.3 $A_s(M)$ and $A_u(M)$ are negatively invariant.

8.6.4 $A_s(M)$ and $A_u(M)$ are invariant whenever π has the openness property (see 1.3.5).

We are now ready for the following basic definition.

8.7 DEFINITION. Let $M \subset X$. Then M is said to be a weak attractor, attractor, strong attractor, uniform attractor if and only if, respectively $A_w(M)$, $A(M)$, $A_s(M)$, $A_u(M)$ is a neighborhood of M.

As an easy but important consequence of the above definition we have

8.8 PROPOSITION. Let $M \subset X$ be a weak attractor, or attractor, or strong attractor, or uniform attractor. Then the corresponding region of attraction $A_w(M)$, or $A(M)$, or $A_s(M)$, or $A_u(M)$ is an open invariant neighborhood of M.

Proof. We first make two observations.

8.8.1 If the net $x\pi t$ is weakly attracted or attracted to a set M, then so is any net $y\pi t$ with $y \in F_x \cup C_x$.

8.8.2 If $y = x\pi t$ for some $t \geqq 0$ and U is a neighborhood of y, then $\pi_t^{-1}(U) = \{z \in X\colon z\pi t \in U\}$ is a neighborhood of x.

We now go back to the proof of our proposition.

8.8.3 Let M be a weak attractor. There is an open neighborhood U of M with $U \subset A_w(M)$. Thus, if for any $x \in X$ and $t \geqq 0$, $x\pi t \in U$, then $x\pi t$ and consequently x is weakly attracted to M (see 8.8.1). However $\pi_t^{-1}(U)$ is a neighborhood of x, and since $y\pi t \in U$ for each $y \in \pi_t^{-1}(U)$, we conclude that $\pi_t^{-1}(U) \subset A_w(M)$. Thus $A_w(M)$ is open.

8.8.4 The proof for $A(M)$ is similar to 8.8.3.

8.8.5 Let M be a strong attractor. We are to show that $A_s(M)$ is open and positively invariant (it is indeed always negatively invariant). Since $A_s(M)$ is a neighborhood of M, $U = \text{Int}\, A_s(M) \supset M$. Let $x \in A_s(M)$. Because x is attracted to M, there is a T, $0 \leqq T < \omega_x$, with $x\pi[T, \omega_x) \subset U$. But then by negative invariance of $A_s(M)$, $x\pi[0,T] \subset F_{x\pi T} \subset A_s(M)$.

Consequently, $C_x = (x\pi[0,T]) \cup (x\pi[T,\omega_x)) \subset A_s(M)$ and $A_s(M)$ is positively invariant. Finally, the set $\pi_T^{-1}(U)$ is an open neighborhood of x and is a subset of $A_s(M)$ as $A_s(M)$ is negatively invariant.

8.8.6 The proof for $A_u(M)$ is analogous to that for $A_s(M)$. Q.E.D.

8.9 COROLLARY. If $M \subset X$ is an attractor, or strong attractor, or uniform attractor, then, respectively, $A(M) = A_w(M)$, or $A_s(M) = A_w(M)$, or $A_u(M) = A_w(M)$.

8.9 together with 8.6.1 show that a uniform attractor is a strong attractor, a strong attractor is an attractor, and that an attractor is a weak attractor. However, the converse implications need not hold in general. We now explore conditions under which converse implications do hold. First the following

8.10 PROPOSITION. $x \in X$ is weakly attracted to $M \subset X$ if and only if either the net $x\pi t$ is frequently in M or $L_x \cap M \neq \Phi$.

Proof.

8.10.1 Let x be weakly attracted to M. Assume that the net $x\pi t$ is not frequently in M, so that it is ultimately in $X - M$. We want to show that $L_x \cap M \neq \Phi$.

Assume the contrary. Then clearly there is a T, $0 \leqq T < \omega_x$, with $K_{x\pi T} = C_{x\pi T} \cup L_x \subset X - M$. Since $K_{x\pi T}$ is closed, $X - K_{x\pi T}$ is a neighborhood of M; but since the net $x\pi t$ is ultimately in $K_{x\pi T}$, it is not frequently in $X - K_{x\pi T}$. This contradicts the fact that the net $x\pi t$ is frequently in every neighborhood of M.

8.10.2 If x is not weakly attracted to M, then there is an open neighborhood U of M with $x\pi t$ ultimately in $X - U$. But then there is a T, $0 \leqq T < \omega_x$, with $x\pi[T, \omega_x) \subset X - U$, so that $K_{x\pi T} \subset X - U$ as the latter set is closed. Thus $L_x \cap M = \Phi$ and $x\pi t$ is not frequently in M. Q.E.D.

If the set M has a compact boundary, then we can assert more.

8.11 THEOREM. Let $M \subset X$ be a non-empty closed set with compact boundary. Then $x \in X$ is weakly attracted to M if and only if either the net $x\pi t$ is ultimately in M or $L_x \cap M \neq \Phi$.

8.12 THEOREM. Let M be a closed subset of X with compact boundary. If $x \in X$ is attracted to M

then either the net $x\pi t$ is ultimately in M or $\Phi \neq L_x \subset M$. The converse holds whenever X is rim-compact.

Proof.

8.12.1 If x is attracted to M and $L_x \neq \Phi$, then $L_x \subset M$. For otherwise, if $M \not\ni y \in L_x$, then there are neighborhoods U of y and V of M with $U \cap V = \Phi$. But then the net $x\pi t$ is frequently in U as $y \in L_x$. This contradicts the fact that the net $x\pi t$ is ultimately in every neighborhood of M, in particular, ultimately in V. If $L_x = \Phi$, and the net $x\pi t$ is frequently in $X - M$, then $x\pi t$ is frequently in ∂M. This follows from the fact that x is weakly attracted to M and therefore $x\pi t$ is also frequently in M as $L_x = \Phi$ (see 8.10). Since ∂M is compact, the net $x\pi t$ has a cluster point in ∂M, which contradicts $L_x = \Phi$. Hence, if $L_x = \Phi$, then the net $x\pi t$ is ultimately in M.

8.12.2 Conversely, assume X is rim-compact. If the net $x\pi t$ is ultimately in M, then it is ultimately in every neighborhood of M and x is attracted to M. If $\Phi \neq L_x \subset M$ then also x is attracted to M. For otherwise there is a neighborhood U of M with compact boundary (X is rim-compact) such that $x\pi t$ is frequently in $X - U$. But then, since x is weakly attracted to M and consequently $x\pi t$ is frequently in U, the net $x\pi t$ is also frequently in the compact set ∂U. Thus the net $x\pi t$ has a cluster point in ∂U, i.e., $L_x \cap \partial U \neq \Phi$. This, however, contradicts $L_x \subset M$. Q.E.D.

8.13 THEOREM. Let π be a global dynamical system in a locally compact space X. Then a set $M \subset X$ is a strong attractor if and only if it is a uniform attractor.

Proof. Let M be a strong attractor. Let $x \in A_s(M)$. Since $A_s(M)$ is open and X locally compact, there is a compact neighborhood U of x with $U \subset A_s(M)$. Let V be an arbitrary neighborhood of M. For each $y \in U$ there is an open neighborhood U_y of y and a $T_y \geqq 0$ such that $U_y\pi[T_y,+\infty) \subset V$. Since $\{U_y\colon y \in U\}$ is an open cover for U, there exist $y_1,y_2,\dots,y_n$ with $P = \cup\{U_{y_i} | i = 1,2,\dots,n\} \supset U$. On setting $T = \max\{T_{y_1},T_{y_2},\dots,T_{y_n}\}$ we get $U\pi[T,+\infty) \subset P\pi[T,+\infty) \subset V$. Since V was an arbitrary neighborhood of M, we have $x \in A_u(M)$ and M is a uniform attractor. The converse always holds. Q.E.D.

We now give some theorems which analyse the effect of stability on attraction properties.

8.14 THEOREM. Let M be a subset of X. If M is orbitally stable and a weak attractor, then M is an attractor and in fact a strong attractor.

Proof. We have to show that $A_w(M) \subset A_s(M)$. Let $x \in A_w(M)$. Let V be any open neighborhood of M. Then by orbital stability of M, there is a positively invariant neighborhood U of M with $U \subset V$ and $U \subset A_w(M)$. Since Int U is also a neighborhood of M, the net $x\pi t$ is frequently in Int U. Thus there is a T, $0 \leqq T < \omega_x$, with $x\pi T \in \text{Int } U$. Choose any neighborhood W of $x\pi T$ with $W \subset \text{Int } U$. Then $W\pi[T,+\infty) \subset U \subset V$ as U is positively invariant. On the other hand, $\pi_T^{-1}(W)$ is a neighborhood of x, and the net $\pi_T^{-1}(W)\pi t$ is ultimately in $U \subset V$. Thus $x \in A_s(M)$. Q.E.D.

8.15 THEOREM. Let $M \subset X$ be closed with compact boundary. If M is a positively invariant strong attractor, then M is orbitally stable.

Proof. If M is not orbitally stable then there is a neighborhood U of M and nets $x_i \to x \in M$,

$t_i \geqq 0$, with $x_i \pi t_i \notin U$. Since M is a strong attractor, there is a neighborhood V of x and a $T \in [0, \omega_x)$ such that $V\pi[T, +\infty) \subset U$. Consequently $0 \leqq t_i \leqq T$. We may therefore assume that $t_i \to t$, $0 \leqq t \leqq T < \omega_x$. But then $x_i \pi t_i \to x\pi t$ as $x\pi t$ is defined. By positive invariance of M, $x\pi t \in M$; on the other hand $x_i \pi t_i \in X - U$ showing that $x\pi t \notin M$. This contradiction proves the theorem. Q.E.D.

We summarize the various implications given in the above theorems in the following diagram. But first the following definition.

8.16 DEFINITION. A set $M \subset X$ is called asymptotically stable if and only if it is an orbitally stable attractor.

8.17 DIAGRAM. Let $M \subset X$

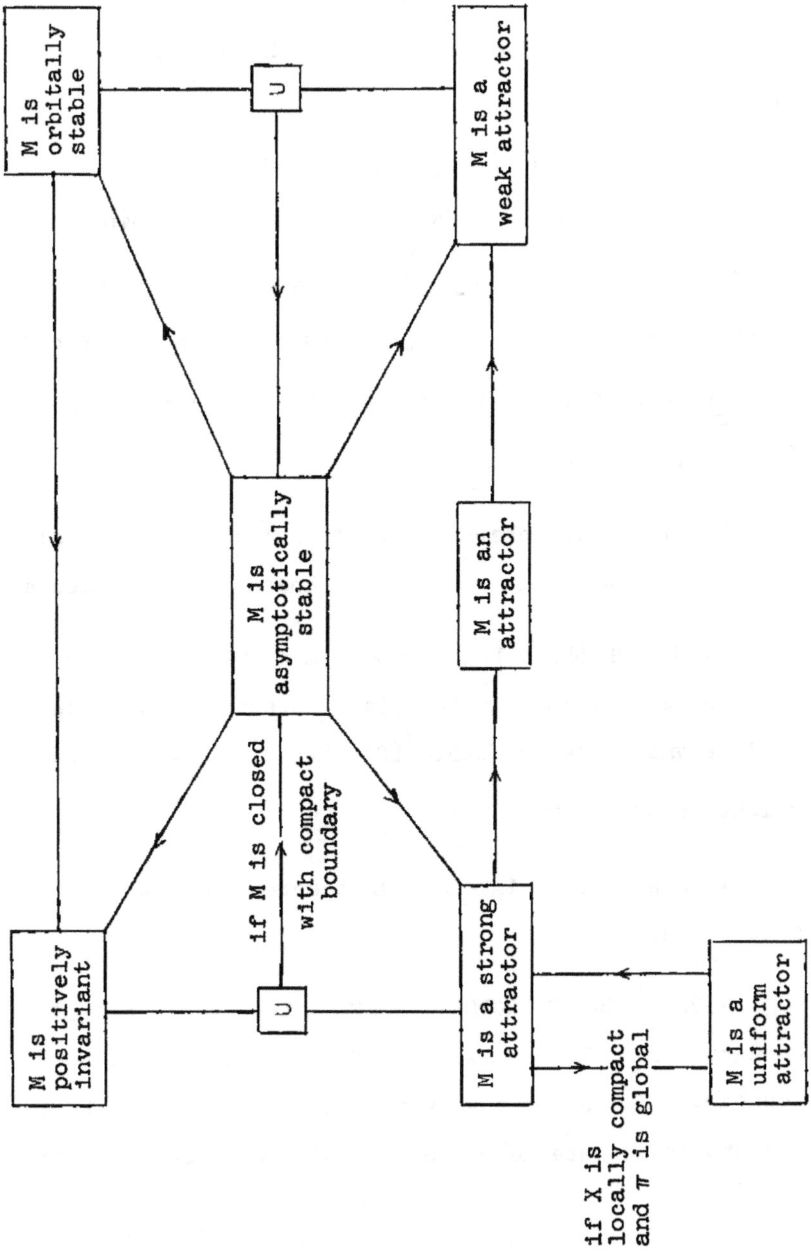

We now give a lemma which leads to a characterization of asymptotic stability.

8.18 LEMMA. For any set $M \subset X$, $x \in A_w(M)$ implies $J_x \subset J(M)$.

Proof. Recall that for any $x \in X$ and t, $0 \leqq t < \omega_x$, $J_x \subset J_{x\pi t}$. Now if $x \in A_w(M)$, then either $x\pi t \in M$ for some t, $0 \leqq t < \omega_x$, or $L_x \cap M \neq \Phi$. In the first case $J_x \subset J_{x\pi t} \subset J(M)$, and in the second case $J_x \subset J_y \subset J(M)$ for an appropriate $y \in L_x \cap M$ (Lemma 6.15). Q.E.D.

The following characterization of an asymptotically stable compact set in rim-compact spaces is now immediate.

8.19 THEOREM. Let X be rim-compact. Let M be a compact positively invariant subset of X. Then M is asymptotically stable iff $\{x: \Phi \neq J_x \subset M\}$ is a neighborhood of M.

We now give an interesting theorem on weak attractors.

8.20 WEAK ATTRACTOR THEOREM. Let X be rim-compact. Let M be a subset of X with compact boundary. If M is a weak attractor then $D(M)$ is an asymptotically stable closed set with compact boundary and $A_w(M) = A_s(M)$. Furthermore, $D(M)$ is the smallest closed orbitally stable set containing M.

Remark. A similar assertion, with the added assumptions that π is a (global bilateral) dynamical system, X locally compact metric, and M a compact strong attractor, was first given in [2]. The significant relaxation of attraction to weak attraction (which e.g. makes possible Theorem 12.8) was given in [3].

For the proof we need the following important

8.21 LEMMA. Under the hypotheses of Theorem 8.20, $D(M)$ is a closed set with compact boundary and $A_w(M)$ is a neighborhood of $D(M)$. In fact $\overline{D(M)-M}$ is compact.

Proof. Let U be a closed neighborhood of M with compact boundary and $U \subset A_w(M)$. For each $x \in \partial U$ define $\tau_x = \inf\{t > 0: x\pi t \in U\}$. Since the net $x\pi t$ is frequently in U, τ_x is well defined and $0 \leqq \tau_x < \omega_x$. We claim that the set

$$P = \cup\{x\pi[0, \tau_x] | x \in \partial U\}$$

is compact. To see this choose any net y_i in P. Then $y_i = x_i\pi t_i$ for some $x_i \in \partial U$ and t_i with $0 \leqq t_i \leqq \tau_{x_i} < \omega_{x_i}$. Indeed, the net x_i has a cluster point in ∂U so that there is a subnet $x_j \to x \in \partial U$.

Since $x \in A_w(M)$, the net $x\pi t$ is frequently in Int U. Thus there is a T, $0 \leqq T < \omega_x$, such that $x\pi T \in \text{Int } U = U^\circ$. But then $\pi_T^{-1}(U^\circ)$ is a neighborhood of x, which shows that $t_j \leqq T$ ultimately. We may therefore assume that $t_j \to \tau$, $0 \leqq \tau \leqq T < \omega_x$. Since $x\pi\tau$ is defined, $x_j\pi t_j \to x\pi\tau$. We need to show that $x\pi\tau \in P$. From the definition of τ_x, for each $x \in \partial U$ we have $x\pi(0,\tau_x) \cap U = \Phi$, $x\pi\tau_x \in \partial U$, and $\partial U \subset P$. Thus if $x\pi\tau \in \partial U$, then indeed $x\pi\tau \in P$. If $x\pi\tau \notin \partial U$, then $x\pi\tau \notin U$, since $x_i\pi t_i \notin \text{Int } U$. Since $x \in \partial U$, there is a t, $0 \leqq t < \tau$, such that $x\pi t \in \partial U$ and $x\pi(t,\tau) \cap U = \Phi$. Then $\tau - t < \tau_{x\pi t}$, so that $x\pi\tau = (x\pi t)\pi(\tau - t) \in P$. We have thus proved that P is compact. Now, the set $U \cup P = V$ is a closed positively invariant subset of the open set $A_w(M)$, and, moreover, V is a neighborhood of M. Therefore $D(M) \subset V \subset A_w(M)$. Since M has compact boundary, $D(M)$ is closed. To prove that $D(M)$ has compact boundary, we show the stronger result that $\overline{D(M)-M}$ is compact, as evidently $\partial(D(M)) \subset \overline{D(M)-M}$. For this choose any net y_i in $\overline{D(M)-M}$, and assume that it has no cluster points. Then there is a closed neighborhood U of M with compact boundary such that some subnet y_j is not in U (otherwise the net y_i is ultimately in every neighborhood of ∂M, and since ∂M is compact, it has a cluster point in ∂M). Since $D(M) \subset U \cup P$ (P constructed for this particular U), we conclude that some subnet y_j is in the compact set P. This contradiction shows that $\overline{D(M)-M}$ is compact. Q.E.D.

Proof of Theorem 8.18. Notice that $D(M) \supset M$ and that $A_w(M)$ is an invariant neighborhood of $D(M)$. Thus clearly $D(M)$ is a weak attractor, and $A_w(D(M)) = A_w(M)$. To see that $D(M)$ is orbitally stable we need only prove $D(D(M)) = D(M)$ (see Lemma 8.21, Theorem 7.4, and Proposition 7.3). For any $x \in D(M) \subset A_w(M)$ we have $J_x \subset J(M)$ by Lemma 8.18. Thus $D(D(M)) = D(M) \cup J(D(M)) \subset D(M) \cup J(M) = D(M)$ where the first equality follows from positive invariance of $D(M)$ and 6.3.1. Thus $D(M)$ is orbitally stable. For the last part of the theorem, note that if $M \subset M^* \subset D(M)$ and M^* is closed, then M^* has a compact boundary ($\partial M^* \subset \overline{D(M)-M}$ which is compact), and

$$D(M) \subset D(M^*) \subset D(D(M)) \subset D(M).$$

Hence $D(M^*) = D(M)$; thus if $M^* \neq D(M)$, then it is not orbitally stable. Q.E.D.

9. FLOW NEAR AN INVARIANT SET

The main purpose of this section is to obtain a generalization of a result due to I. Kimura and T. Ura [14; Proposition 3] which the first author rediscovered using his weak attractor theorem [3]. The following statement on this result was first made in [4].

9.1 THEOREM. Let π be a (local bilateral) dynamical system defined on a locally compact phase space X. Let $M \subset X$ be a compact invariant set. Then one of the following conditions holds:

9.1.1 M is positively asymptotically stable.

9.1.2 M is negatively asymptotically stable.

9.1.3 every neighborhood of M contains a complete trajectory which does not meet M.

9.1.4 there are points $x, y \notin M$ such that $\Phi \neq L_x^+ \subset M \supset L_y^- \neq \Phi$.

In the above statement L_x^+, L_x^- denote respectively, the positive and the negative limit set of an $x \in X$ (see [15], [6]), and positive and negative asymptotic stability is as defined for example in [6]. In the case of the present set up of an lsd-system π on X, it is in general not convenient or desirable to introduce a negative limit set for points $x \in X$, though one may conveniently introduce the negative limit set of a solution σ (see 5.15). We therefore proceed to introduce certain notions which are suited to the aim of this section. We will assume given an lsd-system π on a Hausdorff space X.

9.2 DEFINITION. A set $M \subset X$ is said to be

9.2.1 unstable iff it is not orbitally stable.

9.2.2 completely weakly unstable iff there is a neighborhood U of M such that $X - U$ is a weak attractor and $A_w(X-U) \supset X - M$.

9.2.3 completely unstable iff there is a neighborhood U of M such that $X - U$ is an attractor and $A(X-U) \supset X - M$.

In analogous fashion one may define complete strong instability and complete uniform instability. The importance of the concept of complete weak instability is underscored by the following theorem.

9.3 THEOREM. Let X be rim-compact. Let $M \subset X$ be a closed negatively invariant set with compact boundary. Then M is completely weakly unstable if and only if there is an open neighborhood U of M with compact boundary such that $X - U$ is asymptotically stable with $A(X-U) \supset X - M$.

Proof. The 'if' part of the theorem is trivial. We prove the 'only if' part. Assume that M is completely weakly unstable. Then there is a neighborhood U of M such that $X - U$ is weakly attracting and $A_w(X-U) \supset X - M$. We may even assume (since X is rim-compact) that U is open and ∂U is compact. Then $V = X - U$ is a closed set with compact boundary and is a weak attractor. By the Weak Attractor Theorem 8.20, $D(V)$ is asymptotically stable and has compact boundary, and $A(D(V)) \supset A_w(V) \supset X - M$. It remains to verify that the open set $X - D(V)$ is a neighborhood of M, i.e., that $X - D(V) \supset M$. To see this we note that if W is a closed neighborhood of V with compact boundary and such that $W \cap M = \Phi$, then, as in the proof of Lemma 8.21, $D(V) \subset W \cup P$, where $P = \cup\{x\pi[0,\tau_x] | x \in \partial W\}$ with τ_x defined as in the proof of Lemma 8.21. Now if $D(V) \cap M \neq \Phi$, then clearly $\Phi \neq D(V) \cap M \subset P$. Thus for any $y \in D(V) \cap M$ there is an $x \in \partial W$ and a $t \in (0,\tau_x)$ such that $y = x\pi t$. Since M was assumed negatively invariant, this implies $x \in M$ or $M \cap W \neq \Phi$ which contradicts the choice of M. Q.E.D.

The following result may be proved in a similar fashion.

9.4 THEOREM. Let X be rim-compact, $M \subset X$ a closed positively invariant set with compact boundary. Let M be completely weakly unstable. Then M is negatively invariant.

We are now ready for the main theorem in this section.

9.5 THEOREM. Let X be locally compact. Let M be a closed invariant set with compact boundary. Then one of the following holds:

9.5.1 M is asymptotically stable.

9.5.2 M is completely weakly unstable.

9.5.3 Every neighborhood U of M contains a principal trajectory which does not meet M.

9.5.4 There exist points $x, y \notin M$ such that $\Phi \neq L_x \subset M$, and there is a left maximal solution σ through y with $\Phi \neq L_\sigma^- \subset M$.

Proof. Let neither of 9.5.1, 9.5.2, and 9.5.3 hold; we will prove that then 9.5.4 holds.

9.5.5 Let U be an open neighborhood of M with compact boundary such that $U - M$ contains no principal trajectory. Since M is not completely weakly unstable, there is an $x \in U - M$ such that net $x\pi t$ is ultimately in $U - M$ and $L_x \cap (X-U) = \Phi$ (Proposition 8.10). Since $\overline{U-M}$ is compact, and the net $x\pi t$ ultimately in $U - M$, we have $\Phi \neq L_x \subset U$;

since L_x is compact, it is weakly invariant and contains only principal trajectories (Theorem 5.7). Then $L_x \cap (X-M)$ is weakly invariant by Lemmas 3.3 and 3.8. Thus if $L_x \cap (X-M) \neq \Phi$, then $U - M$ contains a principal trajectory, contradicting our original hypothesis. Thus $\Phi \neq L_x \subset M$.

9.5.6 We now claim that under our assumption, M is not orbitally stable, so that $D(M) \neq M$. For otherwise there exist arbitrarily small positively invariant closed neighborhoods U of M with $\overline{U-M}$ compact. For any such neighborhood U, $x \in U - M$ implies $\Phi \neq L_x \subset \overline{U-M}$ and L_x is compact. If there were a neighborhood U with $\Phi \neq L_x \subset M$ for every $x \in U$, then M would be asymptotically stable, contradicting the assumption that 9.5.1 does not hold. Thus if M is stable, then every neighborhood U of M contains $x \in U - M$ with $\Phi \neq L_x \cap (U-M)$; this is in contradiction with the assumption that 9.5.3 does not hold for M. Consequently M is not orbitally stable.

9.5.7 We now prove the existence of a $y \notin M$ and a left maximal solution σ through y satisfying $\Phi \neq L_\sigma^- \subset M$. Since M is unstable, $D(M) \neq M$. Choose $y \in D(M) - M$, and an arbitrarily closed neighborhood U with compact $\overline{U-M}$, such that $y \notin U$. There is a net $x_i \pi t_i \to y$ with $x_i \to x \in \partial M$ and $t_i \to +\infty$ (indeed, by Lemma 3.4, ∂M is positively invariant, so that compactness of ∂M implies $\omega_x = +\infty$; also $y \in D(M)$, $y \notin M$ shows that $y \in J(M)$ as M is positively invariant). We may assume that $x_i \pi t_i \notin U$, and $x_i \in \text{Int } U - M$. But then there exist τ_i, $0 \leq \tau_i < t_i$, such that the curves $x_i \pi [0, \tau_i] \subset U - M$, and that $x_i \pi \tau_i \in \partial U$. As ∂U is compact, we may assume that $x_i \pi \tau_i \to z \in \partial U$, and also $z \in J_x$. Hence $\tau_i \to +\infty$. Thus indeed the conditions of Lemma 4.6 hold for the compact set $\overline{U-M}$ and the point z. Thus there is a principal negative trajectory through z which remains in $\overline{U-M}$. The corresponding left maximal solution σ through z has $\Phi \neq L_\sigma^- \subset U$, and L_σ^- is compact. L_σ^- is thus weakly invariant and contains only principal trajectories. One now sees easily that $L_\sigma^- \subset M$: otherwise U will contain a principal trajectory not meeting M, thus contradicting the assumption that 9.5.3 does not hold. Q.E.D.

10. LIAPUNOV FUNCTIONS

10.1 DEFINITION. Given, an lsd-system π in a topological space X and a mapping $v: X \to R^1$. Then v is called a Liapunov function (of π, or relative to π) iff it is continuous and

$$v(x\pi t) \leq v(x) \quad \text{for all} \quad x \in X,\ t \in R^+.$$

10.2 The second property may be expressed by saying that v is non-increasing along trajectories of π. In a similar vein we say that v strictly decreases along trajectories iff

$$v(x\pi t) < v(x) \quad \text{for} \quad x \in X,\ t > 0;$$

that it is additive along trajectories (or is a parallelizing function) iff

$$v(x\pi t) = v(x) - t \quad \text{for} \quad x \in X,\ t \in R^+;$$

that it is ultimately constant along a trajectory C_x iff there is a $t_0 \in R^+$ such that

$$v(x\pi t) = v(x\pi t_0) \quad \text{for} \quad t \geq t_0.$$

Given a positively invariant subset Y of X, a Liapunov function of π relativized to Y is called a Liapunov function (of π) on Y.

Evidently, if v is a Liapunov function and

$$v(x\pi t_1) = v(x\pi t_2)$$

for some x, $t_1 \leq t_2$, then also $v(x\pi s) = v(x\pi t_1)$ for all $s \in [t_1, t_2]$. The following lemma exhibits the construction, to a given Liapunov function, of a second Liapunov function which is better in the sense that it strictly decreases as much as possible. This will be needed in the proof of Theorem 10.6.

10.3 LEMMA. Let v be a non-negative Liapunov function of a global semi-dynamical system π on X. Then

10.3.1 $$w(x) = \int_0^{+\infty} e^{-\theta} v(x\pi\theta)d\theta$$

defines a Liapunov function w of π with the following properties: for each $x \in X$,

10.3.2 w strictly decreases on C_x if v is not ultimately constant on C_x;

10.3.3 $w \leq v$;

10.3.4 $\inf w(C_x) = \inf v(C_x)$; in particular, $w \geq 0$ on X.

Proof.

10.3.5 Since π is global, and $v \geq 0$ is non-increasing continuous, 10.3.1 yields a well-defined map $w: X \to R^1$. It is readily verified that w is continuous; e.g., for any $x \in X$ and $\varepsilon > 0$ choose $\tau \in R^+$ with $e^{-\tau}(v(x)+1) < \frac{\varepsilon}{3}$, and then a neighborhood U of x such that, for all $y \in U$, simultaneously

$$v(y) < v(x) + 1,$$

$$|v(y\pi\theta)-v(x\pi\theta)| < \frac{\varepsilon}{3} \quad \text{for all} \quad \theta \in [0,\tau].$$

10.3.6 To show that w non-increases along trajectories, take any $x \in X$, $t \in R^+$. Then

10.3.7 $w(x) - w(x\pi t) =$

$$= \int_0^{+\infty} e^{-\theta}(v(x\pi\theta)-v(x\pi(t+\theta)))d\theta \geq 0,$$

since v is a Liapunov function, so that the integrand in 10.3.7 is non-negative.

10.3.8 To verify 10.3.2 assume $w(x) = w(x\pi t)$, $t > 0$. Then from 10.3.7 again,

$$v(x\pi\theta) - v(x\pi(t+\theta)) = 0 \quad \text{for all} \quad \theta \in R^+;$$

in particular,

$$v(x) = v(x\pi t) = v(x\pi 2t) = \ldots$$

and v is constant on C_x.

10.3.9 Since v non-increases along trajectories,

$$w(x) = \int_0^{+\infty} e^{-\theta} v(x\pi\theta)d\theta \leq \int_0^{+\infty} e^{-\theta} v(x)d\theta = v(x),$$

verifying 10.3.3, and also $\inf w(C_x) \leq \inf v(C_x)$. To complete the proof of 10.3.4, let $v(x\pi t) \geq \mu$ for all $t \in R^+$. Then also $v(x\pi(t+\theta)) \geq \mu$ for all $\theta \in R^+$, and from 10.3.1 directly,

$$w(x\pi t) \geq \int_0^{+\infty} e^{-\theta}\mu d\theta = \mu$$

for all $t \in R^+$. Thus indeed $\inf w(C_x) \geq \inf v(C_x)$ as required. This concludes the proof.

10.4 Remarks. In the case of global dynamical systems more can be proved. One applies the formula

$$w(x) = \int_{-\infty}^{+\infty} e^{-|\theta|} v(x\pi\theta)d\theta$$

starting with a bounded Liapunov function v. This yields a Liapunov function w which has even better properties than that constructed in Lemma 10.3. Namely, w is strictly decreasing along all trajectories on which v is merely non-constant; second, $\lambda > v(x\pi t) > \mu$ for all $t \in R^1$ implies $\lambda > w(x\pi t) > \mu$ for all $t \in R^1$; third, for each $x \in X$,

$$\sup w(C_x) = \sup v(C_x), \qquad \inf w(C_x) = \inf v(C_x).$$

The tan-arctan trick then makes it possible to treat even unbounded Liapunov functions. (In this case we do not have $w \leq v$.)

10.5 The principal result of this section concerns the characterization, under certain circumstances, of asymptotic stability as defined in Section 8 in terms of existence of appropriate Liapunov functions. It seems useful to present separately the necessary and the sufficient conditions: the Stability Theorem is given in 10.10, and 10.6 in the counterpositive formulation may be termed the Instability Theorem. Readers familiar with the proofs of these theorems in other situations will observe that we have assembled a number of known technical devices in order to carry through the proofs in the present rather general situation.

10.6 THEOREM. Assumptions: π is an lsd-system in a locally compact Hausdorff space X; M is closed, invariant and asymptotically stable; ∂M is compact and isolated from start points in $X - M$.

Conclusion: There exists a Liapunov function $v\colon X \to [0,1]$ of π such that

$$v(x) = 0 \quad \text{iff} \quad x \in M, \qquad v(x) = 1 \quad \text{iff} \quad x \notin A(M),$$

v strictly decreases to 0 along trajectories in $A(M) - M$; $\{x: 0 < v(x) < \frac{1}{2}\}$ has compact closure and is within $A(M) - M$.

Proof.

10.6.1 We will actually prove a minor variation of this assertion, where the function value 1 is replaced by $+\infty$; the transition may then be effected via $\xi \to \xi/(1+\xi)$, $\infty \to 1$ in the usual manner.

10.6.2 There exists a closed neighborhood U of M which is a positively invariant subset of $A(M)$, and such that $U - M$ contains no start points and has compact closure.

Indeed, we first construct a compact neighborhood V of the compact set ∂M such that $V \subset A(M)$ and $V - M$ contains no start points. Then, applying orbital stability of M, we take a positively invariant neighborhood W of M within the neighborhood

$$V \cup M \subset A(M);$$

then $W - M$ is positively invariant, without start points, and contained in the set $V - M$ with compact closure; hence $\overline{W-M}$ is positively invariant compact and without start points. Finally set $U = M \cup \overline{W-M}$.

10.6.3 We note two consequences of our assumptions. First, since $U - M$ is positively invariant with compact closure, Theorem 1.11 yields that

$$\omega_x = +\infty \quad \text{for all} \quad x \in U - M.$$

Second, we have

$$J(A(M)) \subset M$$

since $J(A(M)) \subset J(M)$ according to 8.12 and 6.15, and $J(M) \subset D(M) \subset M$ from stability of M (see 7.3).

10.6.4 There exists a continuous function f on X with

$$f\colon X \to [0,1], \quad f|M = 0, \quad f|X - U = 1.$$

Indeed, start with a separating function [13, 5.18] to the compact set ∂M and its neighborhood U, then redefine it as 0 on M. This will not affect the function values on $X - U$, and it will remain continuous since both M and ∂M are closed.

10.6.5 Now define a second function g on X by letting

$$g(x) = \sup f(C_x) \quad \text{for} \quad x \in X.$$

Then, of course,

$$0 \leq f \leq g \leq 1, \quad g|M = 0, \quad g|X - U = 1$$

(for $g|M = 0$ apply invariance of M). Since always $C_{x\pi t} \subset C_x$, g non-increases along trajectories.

10.6.6 Rather obviously, g is lower semi-continuous. Indeed, if $g(x) > \mu$, then $f(x\pi t) > \mu$ for some

$t \in R^+$ by definition of g, whereupon (with t held fixed) also $f(y\pi t) > \mu$ for y near x since f and π are continuous; therefore

$$g(y) \geq f(y\pi t) > \mu$$

for y near x. Thus $\{x: g(x) > \mu\}$ is open, g is lower semi-continuous.

10.6.7 To show that g is actually continuous, assume the contrary: from 10.6.6, that g is not upper semi-continuous. Thus, for some $x_i \to x$ and $\varepsilon > 0$,

$$g(x_i) > g(x) + \varepsilon.$$

Now $0 \leq g \leq 1$, so that necessarily $1 > g(x) \geq f(x)$ and $x \in \text{Int } U$ (since also $f|\overline{X-U} = 1$), whereupon ultimately $x_i \in U$. Secondly, by definition of g, there exist $t_i \in R^+$ with $f(x_i\pi t_i) > g(x) + \varepsilon$. Now either the t_i have a convergent subnet $t_j \to t \in R^+$; but then $x_j\pi t_j \to x\pi t$ and

$$f(x\pi t) \leftarrow f(x_j\pi t_j) > g(x) + \varepsilon \geq f(x\pi t) + \varepsilon,$$

a contradiction with $\varepsilon > 0$. Or $t_i \to +\infty$; in this case

$$f(x_i\pi t_i) > g(x) + \varepsilon \geq \varepsilon > 0$$

so that $x_i\pi t_i$ are not in M (recall that $f|M = 0$), and hence they are in $U - M$ (x_i in U, U positively invariant). Since $\overline{U-M}$ is compact, there is a convergent subnet $x_j\pi t_j \to y$. By definition, then,

$$y \in J_x \subset J(U) \subset J(A(M)) \subset M,$$

so that $f(y) = 0$; again, this contradicts $x_j\pi t_j \to y$, continuity of f and $f(x_i\pi t_i) \geq \varepsilon > 0$. Therefore, finally, g is continuous.

10.6.8 It was already remarked that g non-increases along trajectories, so that it is a Liapunov function.

10.6.9 As the next step, set

$$G = \left\{x: g(x) < \tfrac{1}{2}\right\}.$$

From 10.6.8, G is open positively invariant; and from the known values of g on M and $X - U$,

$$M \subset G \subset \overline{G} \subset \left\{x: g(x) \leq \tfrac{1}{2}\right\} \subset U \subset A(M).$$

Now consider the set $G_t = M \cup (G-M)\pi t$ for $t \in R^+$. (If M were compact without start points, we need have only considered $G\pi t$.) Evidently

$$M \subset G_t \subset G\pi t \subset G.$$

We wish to prove, first, that G_t is a neighborhood of

M (in the case of dynamical systems, $H\pi t$ is open if H is such). Assume not; then there exist $x_i \to x$ with $x \in M$, $x_i \notin G_t$. Since G is an open neighborhood of M, we may as well assume that all $x_i \in G$; then of course

$$x_i \in G - G_t \subset U - M,$$

and the set $U - M$ contains no start points and its closure is compact. Since M is actually strongly negatively invariant, this is possible only if there exist y_i, t_i with

$$y_i \pi t_i = x_i, \quad y_i \in \partial G, \quad 0 \leq t_i \leq t.$$

Now, $\partial G \subset U - M \subset \overline{U-M}$ is compact, so that there exist convergent subnets

$$y_j \to y \in \partial G, \quad t_j \to t' \in [0,t].$$

Then

$$M \ni x \leftarrow x_j = y_j \pi t_j \to y\pi t' \in \partial G\pi[0,t];$$

since M is invariant, this even implies $M \cap \partial G \neq \Phi$, contradicting the fact that G is a neighborhood of M. Thus G_t is a neighborhood of M for each $t \in R^+$.

10.6.10 Now take $t = n = 0,1,2,\ldots$ and prove

$$M = \cap_{n \geq 0} G_n.$$

Since each G_n contains M, we only have to show that $x \in G_n$ for all n implies $x \in M$. Assume not; then $x \in (G-M)\pi n$, so that there exist y_n with

$$x = y_n \pi n, \quad y_n \in G - M.$$

As before, $G - M \subset \overline{U-M}$ is compact, so that there is a convergent subnet $y_i \to y \in \overline{U} = U \subset A(M)$. But then the corresponding subnet $n_i \to +\infty$, so that

$$x \in J_y \subset J(A(M)) \subset M,$$

contradicting the assumption $x \notin M$, and thereby establishing our formula for M.

10.6.11 Since G_n is a neighborhood of M,

$$M \subset \cap \text{ Int } G_n \subset \cap G_n = M$$

and therefore M is a (closed) G_δ set. It follows that M is a zero-set, so that in 10.6.4 we could have taken f such that actually $M = f^{-1}(0)$. Since $0 \leq f \leq g$ in 10.6.5, we even have $M = g^{-1}(0)$; assume this.

10.6.12 Now apply Lemma 10.3 to g, and obtain a Liapunov function w with the properties described there. In particular, $0 \leq w \leq g \leq 1$; $M = w^{-1}(0)$ from $M = g^{-1}(0)$ and invariance of M. We verify that w is strictly decreasing along trajectories in

$A(M) - M$; according to 10.3.1 it suffices to show that g is not ultimately constant along C_x for $x \in A(M) - M$. And indeed, for these x we have $g(x) > 0$ and also that $x\pi t$ is ultimately in every neighborhood of M, in particular in

$$\{x\colon g(x) < \epsilon\}$$

for each $\epsilon > 0$; on the other hand, $g(x\pi t) > 0$ for all $t \in R^1$, from invariance of $M = g^{-1}(0)$.

10.6.13 Now set

$$H = \left\{x\colon w(x) < \tfrac{1}{2}\right\}.$$

Then H is open and

$$M \subset H \subset \overline{H} \subset \overline{G} \subset U \subset A(M)$$

where we have used $w \leq g$ and

$$\overline{G} \subset \left\{x\colon g(x) \leq \tfrac{1}{2}\right\} \subset U.$$

Since w is actually strictly decreasing in $A(M) - M$, we even have

$$\overline{H} = \left\{x\colon w(x) \leq \tfrac{1}{2}\right\}, \qquad \partial H = \left\{x\colon w(x) = \tfrac{1}{2}\right\},$$

so that $\partial H \subset U - M$; and also

$$\partial H\pi(0,+\infty) \subset H.$$

10.6.14 Consider any $x \in A(M) - H$. Then $w(x) \geq \frac{1}{2}$; and, since ultimately $x\pi t \in H$, i.e., $w(x\pi t) < \frac{1}{2}$, there exists a $\tau(x) = \tau_x \in R^+$ with

$$w(x\pi\tau_x) = \frac{1}{2}, \quad \text{i.e.} \quad x\pi\tau_x \in \partial H.$$

Since w strictly decreases in $A(M) - M$, τ_x is determined uniquely by x. In particular, if x and $x\pi t \in A(M) - H$, then

10.6.15 $$\tau(x\pi t) = \tau(x) - t,$$

so that τ strictly decreases along arcs of trajectories entirely in $A(M) - H$.

10.6.16 τ_x depends continuously on $x \in A(M) - H$.

First assume $\tau_x > 0$. Then

$$x\pi(\tau_x - \epsilon) \notin \overline{H}, \quad x\pi(\tau_x + \epsilon) \in H$$

for small $\epsilon > 0$, so that also

$$y\pi(\tau_x - \epsilon) \notin \overline{H}, \quad y\pi(\tau_x + \epsilon) \in H$$

for y near x, i.e., $\tau_x - \epsilon < \tau_y < \tau_x + \epsilon$. Similarly for the case $\tau_x = 0$. Then, for any $\epsilon > 0$ and $y \in A(M) - H$ sufficiently near x, $0 \leq \tau_y < \tau_x + \epsilon = \epsilon$.

10.6.17 Finally, define a map $v: A(M) \to R^+$ by the following assignments:

$$v(x) = \begin{cases} w(x) & \text{if } x \in \overline{H} \\ \frac{1}{2}\exp \tau(x) & \text{if } x \in A(M) - H. \end{cases}$$

This is a well-defined continuous map, since on

$$\overline{H} \cap A(M) - H = \partial H$$

we have

$$w(x) = \frac{1}{2} = \frac{1}{2} e^0 = \frac{1}{2} \exp \tau(x).$$

Furthermore, from the Liapunov properties of w and τ established in 10.6.12 (and 10.6.15), $M = v^{-1}(0)$ and v strictly decreases along trajectories in $A(M) - M$.

10.6.18 To prove that v may be continuously extended (as ∞) to $X - A(M)$ we still have to verify that

$$A(M) \ni x_i \to x \in \partial A(M) \quad \text{implies} \quad v(x_i) \to +\infty.$$

Obviously $x_i \notin \overline{H}$ ultimately, so that we need to prove that $\tau(x_i) \to +\infty$. Assume not. Then $\tau(x_j) \to \tau \in R^+$ for some subnet; since ∂H is compact ($\partial H \subset U - M$, cf. 9.6.2), we may as well assume that $x_j \pi \tau(x_j) \in \partial H$ has a limit, say $y \in \partial H$. Hence

$$x\pi\tau \leftarrow x_j \pi \tau(x_j) \to y.$$

Then $x\pi\tau = y \in \partial H \subset A(M)$; since $A(M)$ is open invariant, this yields the contradiction

$$\partial A(M) \ni x \in A(M).$$

10.6.19 It only remains to prove that v tends to 0 along every C_x with $x \in A(M) - M$. This follows immediately from the facts that v decreases along C_x, and that $x\pi t$ is ultimately in every neighborhood of M, e.g. in

$$\{x: v(x) < \varepsilon\}, \qquad \varepsilon > 0.$$

This concludes the proof of Theorem 10.6.

10.7 COROLLARY. Under the assumptions of Theorem 10.6, M is a closed G_δ set, and $A(M)$ is an open F_σ set.

10.8 Remark and example. The preceding property of asymptotically stable sets should be emphasized. It could not have been noticed in the setting of metric carrier spaces, since there every closed set is a G_δ. On the other hand, even in compact Hausdorff phase spaces, the compact absolutely stable sets need not be G_δ sets, so that their stability behavior cannot be described by a single Liapunov function.

Let X be the set of all ordinals up to and including the first uncountable ordinal ω_1. Endow X with the natural topology (open intervals form an open set base), and with the immobile dynamical system (every point is critical). Then X is compact Hausdorff; every closed subset is absolutely stable, in particular the compact singleton (ω_1) is absolutely stable. But (ω_1) is not the zero set of any continuous function $v: X \to R^1$, in particular, of no Liapunov function.

10.9 COROLLARY. In the same situation, K_x is compact and hence C_x is principal for every $x \in A(M) - M$.

Proof. Either $x \in H - M$, whereupon $C_x \subset H - M \subset U - M$ has compact closure; or $x \in A(M) - H$ and $C_x = x\pi[0,\tau_x] \cup x\pi(\tau_x,+\infty) \subset x\pi[0,\tau_x] \cup H - M$ and this again has compact closure.

10.10 STABILITY THEOREM. Assumptions: π is an lsd-system in a locally compact Hausdorff space X; M is a closed positively invariant subset of X with ∂M compact; $v: G \to R^+$ is a continuous function on a neighborhood G of M with the following properties: $v(x\pi t) \leq v(x)$ whenever $x\pi[0,t] \subset G$, $v(x) = 0$ iff $x \in M$, v is not constant on any $C_x \subset G - M$.

Conclusion: M is asymptotically stable.

Proof.

10.10.1 We may as well assume that G is open. For $\alpha > 0$ define

$$V_\alpha = \{x \in G: v(x) < \alpha\},$$

so that

$$M \subset V_\alpha \subset V_\beta \subset G \quad \text{for} \quad 0 < \alpha < \beta,$$

and V_α is open in G and hence is an open neighborhood of M.

10.10.2 From our assumptions it follows that there exists a (closed) neighborhood U of M with

$$M \subset U = \overline{U} \subset G, \qquad \partial U \subset \overline{U-M} \ \text{compact};$$

in particular, ∂U is a compact subset of $G - M$, so that

$$\inf v(\partial U) = \lambda > 0.$$

10.10.3 Consider the set $U \cap V_\alpha$ for $0 < \alpha \leq \lambda$. By construction, it is a neighborhood of M within G. We wish to prove that $U \cap V_\alpha$ is positively invariant. Assume not; then there exist x and $t \geq 0$ such that

$$x\pi[0,\xi] \subset U \cap V_\alpha$$

for $\xi = t$ but not for any $\xi > t$. Since $U \subset G$ = domain v with G open, $x\pi[0,\xi]$ is within domain v for small $\xi - t > 0$. Hence

$$\alpha > v(x) \geq v(x\pi t) \geq v(x\pi\xi)$$

and therefore also $x\pi\xi \in V_\alpha$ for small $|\xi-t|$. Therefore we must have $x\pi\xi \notin U$ for arbitrarily small $\xi - t > 0$, and thus $x\pi t \in \partial U$. But this implies

$$v(x\pi t) \geq \inf v(\partial U) = \lambda \geq \alpha,$$

contradicting $v(x\pi t) \leq v(x) < \alpha$ again. Therefore indeed $U \cap V_\alpha$ is positively invariant.

10.10.4 To prove that M is (orbitally) stable it suffices to show that every neighborhood V of M contains some one of the positively invariant neighborhoods $U \cap V_\alpha$ constructed in 10.10.3. Assume the contrary, that for each α, $0 < \alpha \leq \lambda$, there exist

$$x_\alpha \in U \cap V_\alpha - V.$$

Since $V \supset M$ and $\overline{U-M}$ is compact, there exist convergent nets

$$x \leftarrow x_i \in U \cap V_{\alpha_i} - V, \quad \alpha_i \to 0.$$

It follows that, first, $x \in \overline{U} \subset G$,

$$0 \leq v(x) \leftarrow v(x_i) < \alpha_i \rightarrow 0, \qquad v(x) = 0$$

and therefore $x \in M$ by assumption; and second, since $x_i \notin V$ with V a neighborhood of M, that $x \notin M$: contradiction. Thus the $U \cap V_\alpha$ with $\alpha > 0$ are arbitrarily small positively invariant neighborhoods of M, and M is orbitally stable.

10.10.5 Finally we prove that every $x \in U \cap V_\lambda$ has $x\pi t$ ultimately in every $U \cap V_\alpha$, $\alpha > 0$. Since the $U \cap V_\alpha$ are arbitrarily small neighborhoods of M, this will verify that $U \cap V_\lambda \subset A(M)$, i.e., that M is an attractor, since $U \cap V_\lambda$ is a neighborhood of M.

Assume the contrary, that there exist

$$\alpha \in U \cap V_\lambda, \quad t_i \rightarrow +\infty, \quad \alpha \in (0, \lambda], \quad x\pi t_i \notin U \cap V_\alpha.$$

Since $U \cap V_\lambda \subset G$ with v non-increasing, this yields that $v(x\pi t) > \alpha$ for all $t \in R^+$. In particular, $C_x \subset U - M$ with $\overline{U-M}$ compact, so that there is a convergent subnet $x\pi t_j \rightarrow y$, and

$$0 < \alpha \leq v(y) \leftarrow v(x\pi t_j), \qquad y \in \overline{U-M} \subset G.$$

For every $t \in R^+$ and t_j there is a $t_k \geq t_j + t$; thus

$$v(x\pi t_k) \leq ((x\pi t_j)\pi t)$$

and, on taking limits,

$$v(y) \leq v(y\pi t)$$

for every $t \in R^+$. Since also $v(y\pi t) \leq v(y)$ and $y \notin M$, this contradicts the assumption that v is non-constant on every trajectory in $G - M$. Q.E.D.

11. THE START POINT SET

By Definition 1.6, a point x in the phase space X of an lsd-system π is a start point of π iff never $x = y\pi t$ for $t > 0$ and $y \in X$. Thus

11.1 LEMMA. The set S of all start points satisfies

$$S = X - X\pi(0,+\infty).$$

11.2 The purpose of this section is to present some result concerning the start point set; the main ones are Theorems 11.8 and 11.11.

In the absence of over-riding remarks we assume given an lsd-system π on a topological space X; the set of all start points of π will be consistently denoted by S as in 11.1. The first two lemmas collect some elementary properties of S.

11.3 LEMMA. S contains no interior points, and is strongly negatively invariant. S contains no periodic nor critical points.
(Proof: On every positive trajectory, all points except at most one are non-start points; hence $\text{Int}\, S = \Phi$. The remaining assertions follow from 11.1.)

11.4 LEMMA. If π has the openness property, then S is closed. If furthermore π also has negative unicity, then S is empty.

Proof. For the first assertion merely observe that, under our assumption, each summand in

$$X\pi(0,+\infty) = \bigcup_{t>0} X\pi t$$

is open.

For the second, take any $x \in X$. Then from axioms 1.1.1 and 1.1.5, there exists $\varepsilon > 0$ and an open set $G \ni x$ such that $y\pi t$ is defined for all $y \in G$, $t \in [0,\varepsilon]$. The assignment $y \to y\pi\varepsilon$ takes G into $G\pi\varepsilon$, an open neighborhood of $x\pi\varepsilon$ by assumption. Hence from continuity, $x\pi(\varepsilon-\delta)$ is in $G\pi\varepsilon$ for small $\delta > 0$; thus there exists $y \in G$ with $x\pi(\varepsilon-\delta) = y\pi\varepsilon$. From negative unicity, then,

$$x = y\pi(\varepsilon-\varepsilon+\delta) = y\pi\delta.$$

Since $\delta > 0$, this shows that an arbitrary $x \in X$ is not a start point, as asserted. Q.E.D.

11.5 Remark. In point of fact, negative unicity and the openness property together form a necessary and sufficient condition for π to be the restriction of a (bilateral) local dynamical system on X; and it is easily verified that such systems cannot have start points (see [10] IV, 2.5 and II, 2.5). In this connection it seems appropriate to mention two other results of [10] (VI, 4.1 and VI, 4.4). Let π be an lsd-system with negative unicity on X. If X is a manifold, then π has the openness property; if X is a manifold-with-boundary, then the boundary of X is strongly negatively invariant.

In the sequel we shall largely attempt to avoid the assumptions of openness and negative unicity. We shall begin the study with several examples.

11.6 Examples.

11.6.1 Consider the so-called additive lsd-system π on R^+: $x\pi t = x + t$ for $x, t \in R^+$. There is a unique start point, at the origin. One may observe that the system is global with negative unicity (so that it illustrates one of the theorems formulated in 11.5). According to 11.4, the system does not have the openness property; this is also easily verified directly. It will be seen from 11.11 (and also from 11.6.3 and 11.6.4) that this situation is rather exceptional: usually the start points cannot be isolated.

11.6.2 The phase space X is a subset of R^2, consisting of the closed right half-plane together with the x-axis. The lsd-system is defined by parallel uniform movement to the right:

$$(x,y)\pi t = (x+t,y) \quad \text{for} \quad (x,y) \in X, \quad t \in R^+.$$

The start point set is the y-axis less the origin (and hence is not closed in X). The lsd-system is global with negative unicity, and hence, again, cannot have the openness property.

11.6.3 Assume given an infinite system of lsd-systems π_i on phase spaces X_i, subject to the condition that infinitely many of them do not have negative globality (see 1.3.2) but otherwise quite arbitrary. We shall consider the product lsd-system π in the sense of [10 , IV, 3.4], acting on the product space $X = \Pi X_i$; and show that the start point set is necessarily dense in X (this result is due to P. N. Bajaj [17])

We may assume that the index set I contains the set of positive integers N, and that for $n \in N$, π_n does not have negative globality. Thus there exist $u_n \in X_n$ with $\alpha_{u_n} > -\infty$, and hence also $v_n \in X_n$ with $\alpha_{v_n} > -\frac{1}{n}$ (see 2.13). It is readily verified that any point

$$x = (x_i) \in \Pi X_i = X$$

such that $x_n = v_n$ for infinitely many $n \in N$ has $\alpha_x = 0$ and hence is a start point for π; and also that the set of such points x is dense in X.

11.6.4 Consider any lsd-system π defined by a functional-differential equation

$$\frac{dx}{dt} = f(\rho_t x)$$

as described in the Section 13. The phase space is the Banach space $C[-1,0]$ of continuous functions $x: [-1,0] \to R$. It is easily seen that if $x = y\pi\epsilon$ for some $y \in C[-1,0]$ and $0 < \epsilon < 1$, then x must be differentiable on $(-\epsilon,0]$ at least. Thus every $x \in C[-1,0]$ which is not differentiable on any $(-\epsilon,0]$ for small $\epsilon > 0$ is necessarily a start point; and therefore the start point set is dense in the phase space.

11.7 Remark. In both of the last two examples the phase space was, essentially, infinite dimensional. The following theorem suggests that this is not accidental. The assertion was conjectured by the first author in the course of a seminar in the spring of 1967. The beautiful proof given below was provided by H. Halkin in a conversation during the 1967 summer school at Varenna.

11.8 THEOREM. An lsd-system on a manifold has no start points.

Proof. Let π be an lsd-system on an n-manifold X; it is required to show that every x_0 is a non-start point, i.e. that

11.8.1 $x_0 = y\pi t$ for some $y \in X$, $t > 0$.

By assumption, x_0 has an open neighborhood homeomorphic to R^n; let us actually identify it with R^n.

Since $x_0\pi t_0$ is defined for small $t_0 > 0$, there is a closed ball E^n centered at x_0 such that $x\pi t$ is defined for all $x \in E^n$, $t \in [0,t_0]$. In particular, a map $f_t\colon E^n \to R^n$ is well-defined by

$$f_t(x) = x_0 + x - x\pi t$$

for $t \in [0,t_0]$, and of course f_t is continuous.

Next, since E^n is compact and π is continuous with the initial value property 1.1.2, for any $\epsilon > 0$ and sufficiently small $t > 0$ one has $\|x\pi t-x\| < \epsilon$ throughout E^n. Now take $\epsilon = \operatorname{diam} E^n$ and fix any corresponding $t > 0$; thus

$$\|x\pi t-x\| < \operatorname{diam} E^n \quad \text{for all} \quad x \in E^n$$

and our particular $t > 0$. This ensures that f_t actually maps into E^n:

$$\|f_t(x)-x_0\| = \|x-x\pi t\| < \operatorname{diam} E^n.$$

By the Brouwer Theorem, the continuous map f_t of the closed ball E^n into itself has a fixed point y;

therefore

$$y = f_t(y) = x_0 + y - y\pi t,$$

which yields 11.8.1 as required.

The next group of results on the start point set concerns the case that the phase space is locally compact Hausdorff. We begin with two lemmas.

11.9 LEMMA. If M is compact and $M \cap M\pi t = \Phi$, then

$$M \subset ((S \cap M) \cup \partial M)\pi[0,t].$$

Proof. Assume that the conclusion does not hold. Then there exists a left maximal solution s through some point $x \in M$ (refer to Section 2) such that

11.9.1 $$s[-t,0] \subset \text{Int } M - S.$$

There are now two cases. Either $-t \in$ domain s; but then $s(-t) \in \overline{\text{Int } M} \subset M$ and

$$s(-t)\pi t = s(0) = x \in M,$$

contradicting $M \cap M\pi t = \Phi$. Or $-t \notin$ domain s, so that $\alpha_s \geq -\epsilon$ and $s[-t,0] = s[-\alpha_s,0]$ is a negative trajectory N in Int M. Since M is compact, according to 2.10 we have that either N is principal, contradicting $\alpha_s \geq -t$; or $s(-\alpha_s)$ is a start point, contradicting 11.9.1. This completes the proof.

11.10 LEMMA. Let π be an lsd-system in a locally compact Hausdorff space. Then, for any open subset G,

11.10.1 $$S \cap G \subset \bigcap_{\epsilon>0} \operatorname{Int}((S \cap G)\pi[0,\epsilon]);$$

in particular,

11.10.2 $$S \subset \bigcap_{\epsilon>0} \operatorname{Int}(S\pi[0,\epsilon]).$$

Proof.

11.10.3 We shall prove 11.10.1 by showing that any point $x \in G$ not in the exhibited intersection is a non-start point. Thus, let

11.10.4 $$x \notin \operatorname{Int}((S \cap G)\pi[0,\epsilon])$$

for some $\epsilon > 0$. This then also holds for smaller $\epsilon > 0$, so we may assume that $x\pi\epsilon$ is defined. Next, there is nothing left to prove if $x = x\pi\epsilon$ (see 11.3); thus we shall also assume $x \neq x\pi\epsilon$.

11.10.5 From our assumptions on the phase space (and the axioms in 1.1) it follows that there exists a compact neighborhood U of x such that

11.10.6 $$U \subset ((S \cap U) \cup \partial U)\pi[0,\epsilon] \subset (S \cap G)\pi[0,\epsilon] \cup \partial U\pi[0,\epsilon].$$

But we also have 11.10.4, so that there exist $x_i \to x$

with $x_i \notin (S \cap G)\pi[0,\epsilon]$. Since U is a neighborhood of x, ultimately $x_i \in U$; hence necessarily $x_i \in \partial U\pi[0,\epsilon]$. In other words, $x_i = y_i\pi t_i$ for appropriate $(y_i, t_i) \in \partial U \times [0,\epsilon]$. The latter set is compact, so there exists a convergent subnet $(y_i, t_i) \to (y,t) \in \partial U \times [0,\epsilon]$. Also, by construction of U, $y\pi t$ is defined, and thus

$$y\pi t \leftarrow y_i\pi t_i = x_i \to x, \quad y\pi t = x.$$

Here one cannot have $t = 0$, since y is on the boundary of a neighborhood of x. Thus, finally, $t > 0$, showing that x is indeed not a start point. This concludes the proof.

11.11 THEOREM. Let π be an lsd-system in a locally compact Hausdorff space X, and assume that every curve $x\pi[0,\epsilon]$ is nowhere dense in X. Then every neighborhood of a start point contains uncountably many start points.

Proof. Assume the contrary, that some $x \in S$ has an open neighborhood G such that $S \cap G$ is countable. Since by assumption each $y\pi[0,\epsilon]$ is nowhere dense,

$$(S \cap G)\pi[0,\epsilon] = \bigcup_y y\pi[0,\epsilon]$$

is meager; from the Baire Theorem, then,

$$\mathrm{Int}((S \cap G)\pi[0,\epsilon]) = \Phi.$$

But then in 11.10.1 the right side is void, while the left side contains x. This contradiction completes the proof.

11.12 Remarks to formula 11.10.2.

11.12.1 It may be observed that $S\pi[0,\epsilon]$ can be written as

$$S\pi[0,\epsilon] = S \cup S\pi(0,\epsilon]$$

where the first summand has a void interior, the second is contained in $X\pi(0,+\infty)$ and hence is disjoint with S (see 11.3 and 11.1).

In locally compact Hausdorff phase spaces,

$$\overline{M} = \bigcap_{\epsilon>0} \overline{M}\pi[0,\epsilon] = \bigcap_{\epsilon>0} \overline{M\pi[0,\epsilon]}$$

for any subset M; in particular, the right hand side in 11.10.2 is contained in $\overline{S}$.

11.12.2 If π has negative unicity (and X is locally compact Hausdorff), then actually

$$S = \bigcap_{\epsilon>0} S\pi[0,\epsilon] = \bigcap_{\epsilon>0} \mathrm{Int}(S\pi[0,\epsilon]),$$

so that S is a G_δ set.

Having 11.10, to verify this we need only show that $\bigcap_{\epsilon>0} S\pi[0,\epsilon] \subset S$. And indeed, if $x \notin S$, there is

$x = y\pi t$ for some $y \in X$, $t > 0$; from negative unicity then, $x \notin S\pi[0, \epsilon]$ for all $\epsilon < t$.

11.12.3 If negative unicity is not assumed, we may have

$$S \neq \bigcap_{\epsilon>0} \mathrm{Int}(S\pi[0,\epsilon]).$$

Indeed, consider the lsd-system suggested by the diagram 11.12.4 (parametrized by arc length). Apparently every $S\pi[0,\epsilon]$ is a neighborhood of x, so that

$$\bigcap_{\epsilon>0} \mathrm{Int}(S\pi[0,\epsilon]) \ni x \notin S.$$

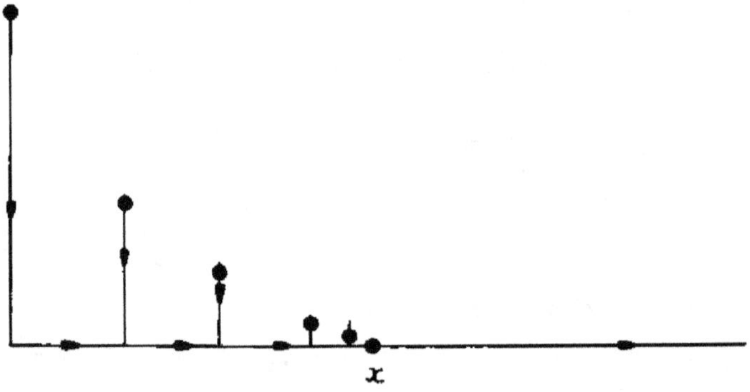

Diagram 11.12.4

11.12.5 In Theorem 11.11 local compactness plays an essential role. Indeed, consider first the set Y of all ordinals $\alpha \leq \omega_1$, and then the subset X of $Y \times R$ obtained by deleting the (strictly) negative axis through ω_1:

$$X = \{(\alpha, s) \in Y \times R\colon \alpha < \omega_1 \quad \text{or} \quad s \geq 0\}.$$

Endow X with the obvious topology, and with the rather natural lsd-system π, defined by

$$(\alpha, s)\pi t = (\alpha, s+t)$$

for $(\alpha, s) \in X$, $t \in R^+$.

Evidently π is a global lsd-system, with negative unicity; and $(\omega_1, 0)$ is the unique start point. Furthermore, S is not a G_δ set, and each $\mathrm{Int}(S\pi[0, \epsilon])$ is void.

12. MINIMALITY; CHARACTERISTIC 0

In this section we first study three versions of minimality (for the corresponding definitions of invariance refer to Section 3). The principal result is 12.9. The results obtained here are then applied to the study of lsd-systems with characteristic 0. Throughout we assume given an lsd-system π on a Hausdorff space X.

12.1 DEFINITION. A subset M of the phase space X is called positively (or weakly, or strongly) minimal iff it is closed and positively (or weakly or strongly, respectively) invariant, but none of its nonvoid proper subsets has these two properties.

12.2 Remarks, examples.

12.2.1 Of course, for weakly invariant sets, positive minimality implies weak minimality; and for strongly invariant sets, weak minimality implies strong minimality. In particular both implications apply if $M = X$. However, in some situations more can be said: see 12.4. It should be noted that actually strong minimality is, generally speaking, the weakest of the three properties.

12.2.2 Obviously a closed positively invariant set M is positively minimal iff every trajectory is dense in M, i.e., iff $K_x = M$ for every $x \in M$. Similar characterizations can be given of weak and strong minimality, using the concept of complete trajectories from 2.12.

12.2.3 Evidently the phase space X is positively minimal iff every open strongly negatively invariant subset M is trivial in the sense that $M = \Phi$ or $M = X$. An analogous characterization can be given of strong minimality.

12.2.4 For (bilateral) dynamical systems, weak and strong minimality coincide. Now consider the additive dynamical system on R. Then R is (strongly) minimal but not positively minimal; and actually the only positively minimal subset is Φ.

12.2.5 Two positively minimal subsets of X either coincide or are disjoint; and similarly for strongly minimal subsets.

12.2.6 In the lsd-system on the triad (see 4.8.1) it is simple to find distinct weakly minimal subsets which do intersect; neither is strongly minimal.

12.3 LEMMA. A subset $M \subset X$ is positively minimal if and only if

12.3.1 $\qquad M = L_x$ for each $x \in M$.

In the positive case $\omega_x = +\infty$ for all $x \in M$.

Proof. Everything is clear if $M = \Phi$; thus we shall assume $M \neq \Phi$.

First let 12.3.1 hold. Then M is closed and positively invariant according to 5.4. Hence $K_x \subset M$ for every $x \in M$, and from 12.3.1 and 5.4.1,

$$M = L_x \subset K_x \subset M;$$

thus $K_x = M$ for every $x \in M$, so that M is positively minimal.

Second, assume that M is positively minimal, but 12.3.1 does not hold, i.e. some $x \in M$ has $L_x \neq M$. In any case $L_x \subset K_x \subset M$ and L_x is closed positively invariant, so that positive minimality then implies $L_x = \Phi$. We conclude, first,

$$M = K_x = C_x \cup L_x = C_x;$$

second, that C_x contains no periodic points; and third, that ${}_x\pi$ maps $[0, \omega_x)$ homeomorphically onto C_x. But then (for $0 < \epsilon < \omega_x$), $x\pi[\epsilon, \omega_x)$ is a closed positively invariant subset of M which is neither empty nor coincides with $C_x = M$. This contradicts positive minimality of M.

The last assertion follows from 5.2.1. Q.E.D.

In Section 5 there were given sufficient conditions for L_x to be weakly invariant. Hence and from Lemma 12.3 we then conclude

12.4 COROLLARY. If M is positively minimal, and either M is compact or X locally compact, then M is weakly minimal without start points.

12.5 THEOREM. Every compact non-void positively invariant subset of the phase space contains a non-void positively minimal set, which is then necessarily compact, weakly minimal and without start points.

Proof. This follows directly by Zorn's Lemma and 3.3.2; the last assertion is a consequence of 12.4. Q.E.D.

12.6 COROLLARY. Every compact weakly minimal set is positively minimal.

Proof. Indeed, let M be compact weakly minimal; we may assume $M \neq \Phi$. According to 12.5, there exists a non-void $M' \subset M$ which is both positively and weakly minimal. But then M' is closed weakly invariant, so that $M' = M$; and therefore $M = M'$ is positively minimal. Q.E.D.

12.7 THEOREM. Every compact non-void strongly invariant subset of the phase space contains a non-void strongly minimal subset.

Proof. Zorn's Lemma and 3.3.2, 3.1.2, 3.1.3. Q.E.D.

In 12.4 and 12.6 we have shown that, for compact sets, positive and weak minimality are equivalent. An application of Theorem 8.20 yields that in some cases compactness need not be assumed, but actually follows.

12.8 THEOREM. In a locally compact phase space, every positively minimal set is compact.

Proof. Let M be a positively minimal subset of a locally compact phase space X. Then M is closed, and hence locally compact; and also positively minimal, so that the restriction

$$\pi' | M \times R^+ : M \times R^+ \to M$$

is an lsd-system on M as phase space, relative to which M is still positively minimal.

Thus we may and shall assume that the original phase space X itself is positively minimal; and we wish to show that X is compact. Of course, we need only consider the case $X \neq \phi$; choose a fixed $p \in X$. From positive minimality and Lemma 12.3,

$$p \in X = L_x \quad \text{for all} \quad x \in X.$$

Thus the singleton $\{p\}$ is a (compact) weak attractor; according to Theorem 8.20, D_p is compact. But, from 12.3 again,

$$X = L_p \subset D_p \subset X;$$

thus $X = D_p$ is compact, as it was required to show.
Q.E.D.

12.9 COROLLARY. In a locally compact phase space, a weakly minimal subset is positively minimal if and only if it is compact.

Proof. 12.8 and 12.6. Q.E.D.

12.10 We shall now study lsd-systems with the property that always $K_x = D_x$. This problem, for global dynamical systems, was suggested by T. Ura, and treated in [1]; it may be noted that they are a generalization of the (positively) Liapunov stable systems [15]. Theorem 12.14 is then the counterpart to Bebutov's theorem on the classification of Liapunov stable systems.

12.11 DEFINITION. An lsd-system π on X is said to have characteristic 0 iff $K_x = D_x$ for all $x \in X$.

12.12 LEMMA. Each of the following conditions is equivalent to π having characteristic 0:

12.12.1 $L_x = J_x$ for every $x \in X$.

12.12.2 Every closed positively invariant set is stable.

Proof. That 12.12.1 is sufficient follows from 5.4.1 and 6.3.1; that it is necessary follows from 5.3 and 6.2.2. Directly from 7.3, 12.12.2 is necessary and sufficient. Q.E.D.

12.13 PROPOSITION. For systems of characteristic 0, each L_x and also each L_σ^- (see 5.15) is positively minimal. Furthermore, if $L_\sigma^- \neq \Phi$, then $K_x = L_\sigma^-$ for $x = \sigma(0)$.

Proof. We will verify 12.3 for L_x. Thus, take any $y \in L_x$; then $L_y \subset L_x$ since L_x is closed positively minimal. Furthermore, from 6.15 (and 12.12.1),

$$L_x = J_x \subset J_y = L_y.$$

Thus we have shown that $L_y = L_x$ for all $y \in L_x$.

The proof for L_σ^- is similar, and uses 6.18 instead of 6.15. Furthermore, then $x \in L_\sigma^-$, hence $K_x \subset L_\sigma^-$, and equality follows from positive minimality. Q.E.D.

12.14 THEOREM. Let π be an lsd-system of characteristic 0, in a locally compact phase space X; and define

$$M = \{x: x \in L_x\}, \quad A = \{x: L_x \neq \Phi\}.$$

Then M is closed and A open; furthermore, $A \supset M$ and $L_x \subset M$ for all $x \in A$.

Proof. Let $x \leftarrow x_1 \in M$. Then $x_1 \in L(x_1) = J(x_1)$, so that $x \in J(x) = L(x)$ (we have applied 12.12.1 and 6.19); hence $x \in M$. This shows that M is closed. Obviously $A \supset M$.

Next we show that $L_x \subset M$ for $x \in A$. From 12.13, L_x is positively minimal; hence and from 12.3, for every $y \in L_x$, we have

$$L_y = L_x \ni y$$

and hence $y \in M$.

Finally we prove that A is open. Take any $x \in A$, so that L_x is closed and positively invariant, and hence stable according to 12.12.2. Secondly, L_x is positively minimal and hence compact (see 12.13 and 12.8). Hence and from 7.4, L_x is orbitally stable. Thus if we take any compact neighborhood U of L_x, there exists a positively invariant neighborhood $V \subset U$

of L_x. Ultimately $x\pi t \in V$; for this t and y sufficiently near x, we have $y\pi t \in V$; hence $C(y\pi t)$ is then within the compact set U. Thus, finally, $L_y = L_{y\pi t} \neq \Phi$ from 5.5. This shows that $y \in A$ for all y near x, as required.

12.15 Remarks to 12.14. It is easy to verify that M is precisely the union of all (compact) positively minimal subsets of X. Evidently $J_x = \Phi$ iff $x \notin A$. Thus, if we use the terminology taken from dynamical system theory, M consists of the Poisson stable points, A is the region of attraction to M and consists of the Lagrange stable points, $A - M$ consists of the asymptotic points, and in $X - M$ the system is dispersive.

13. FUNCTIONAL-DIFFERENTIAL EQUATIONS

13.1 In this section we shall be concerned with functional-differential equations with bounded time-lag, which we shall denote by

13.2 $$\frac{d}{dt}\, x(t) = f(\rho_t x)$$

for given $f\colon C[-1,0] \to R^n$; here $\rho_t x$ is the "t-translation" of x.

13.3 To make this precise, we must be more specific. To this end, in the present section we shall use the notation and terminology described below. It will be seen that little generality is lost by considering R instead of R^n, i.e., by considering one-dimensional systems (or, first-order equations).

13.3.1 P consists of all continuous functions $x\colon (-\infty, \omega_x) \to R$ where $\omega_x \in R$ or $\omega_x = +\infty$.

13.3.2 For $t \in R$, ρ_t denotes the partial mapping from P into $C[-1,0]$ defined by letting

$$(\rho_t x)(s) = x(t+s) \quad \text{for} \quad s \in [-1,0]$$

whenever possible, i.e. iff $t < \omega_x$.

13.3.3 f denotes an arbitrarily fixed mapping $C[-1,0] \to R$ which (with the uniform topology for $C[-1,0]$ -- this is implicit in the entire exposition) is continuous and bounded on bounded subsets of $C[-1,0]$. This class includes all Lipschitzian mappings f, e.g.

$$f(x) = \|a \cdot x+b\| = \sup_{[-1,0]} |a(s)x(s)+b(s)|$$

for given a, b in $C[-1,0]$ and all $x \in C[-1,0]$. Thus it also includes all linear functionals on $C[-1,0]$, and in particular maps of the form

$$f(x) = \Sigma_1^n \alpha_k x(s_k) + \int_{-1}^{0} x(s)da(s)$$

for given $\alpha_k \in R$, $s_k \in [-1,0]$, absolutely continuous a, and all $x \in C[-1,0]$.

13.3.4 We shall say that x is a solution of 13.2 iff $\omega_x > 0$ and 13.2 holds for all $t \in (0,\omega_x)$. In the positive case, the initial value of x is defined to be $\rho_0 x$, i.e. as that element $x_0 \in C[-1,0]$ for which

$$x_0(s) = x(s) \quad \text{for} \quad s \in [-1,0].$$

13.3.5 We shall say that 13.2 has positive uniqueness (of the forward initial value problem) iff the following condition holds: Whenever x, y are solutions of 13.2 with coinciding initial values, i.e. with

$x(s) = y(s)$ for $s \in [-1,0]$, then also $x(s) = y(s)$ even for all s with $-1 \leq s < \min \omega_x, \omega_y$.

It can be shown that, as in the case of ordinary differential equations, this condition is automatically satisfied for all Lipschitzian f, but is not satisfied for all mappings f described in 13.3.3 (e.g., for f defined by $f(x) = {}^{3/2}\sqrt{|x(0)|}$).

13.3.6 Assume that positive uniqueness obtains. Then one may associate with 13.2, in a significant manner, another type of object. Take any $x \in C[-1,0]$, and construct (if possible) a solution y of 13.2 with initial value x. Then define

$$x\pi t = \rho_t y$$

for $0 \leq t < \omega_y$. Positive uniqueness ensures that $x\pi t$ is independent of the particular choice of y, so that π is a partial map from $C[-1,0] \times R^+$ into $C[-1,0]$.

This is illustrated in fig. 13.3.7 for the case $dx/dt = x(t-1)$ and one choice of initial value.

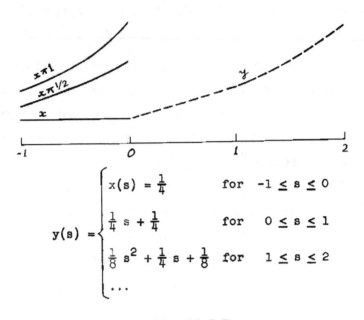

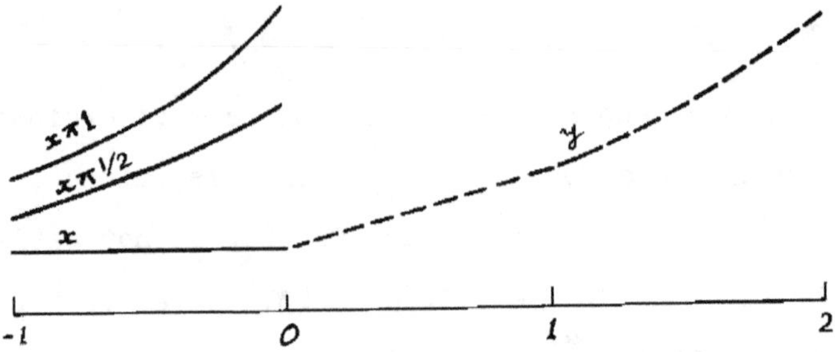

$$y(s) = \begin{cases} x(s) = \frac{1}{4} & \text{for } -1 \leq s \leq 0 \\ \frac{1}{4}s + \frac{1}{4} & \text{for } 0 \leq s \leq 1 \\ \frac{1}{8}s^2 + \frac{1}{4}s + \frac{1}{8} & \text{for } 1 \leq s \leq 2 \\ \cdots \end{cases}$$

Fig. 13.3.7

We are now ready to formulate the principal result of this section.

13.4 THEOREM. If $f\colon C[-1,0] \to R$ is continuous and bounded on bounded subsets, and if positive uniqueness obtains, then the partial map π defined in 13.3.6 is an lsd-system on $C[-1,0]$; furthermore, each right solution ${}_x\pi$ of π (see 1.12.1) is a solution of 13.2 with initial value x.

13.5 The last assertion, and also the initial value and semi-group axioms, follow directly from 13.3.6; the verification of the local existence, continuity and

Kamke's axioms is more involved. However, a large portion of this has been performed in a considerably more general situation [9 , Prop. 9]; and all it remains to verify is the following condition

13.5.1 Under the assumptions of 13.4, for every $x \in C[-1,0]$ there exists $\epsilon > 0$ such that $y\pi t$ is defined and continuous for $\|y-x\| < \epsilon$, $0 \leq t < \epsilon$.

This will be proved via one construction and a series of lemmas. Most of the latter are independent results, and do not require the positive uniqueness condition.

13.6 We begin with a continuity property of the t-translation ρ_t; the entirely straightforward proof is omitted.

LEMMA. For any x_i and x in P, t_i and t in R such that

$$t_i \to t < \omega_x \leq \liminf \omega_{x_i}$$

and that $x_i \to x$ uniformly on compact subsets of $(-\infty, \omega_x)$, we have that $\rho_{t_i} x_i \to \rho_t x$ in $C[-1,0]$.

13.7 COROLLARY. $x \in P$ is a solution of 13.2 if and only if $\omega_x > 0$ and

13.7.1 $$x(s) = x(0) + \int_0^s f(\rho_t x)dt$$

for all $s \in [0, \omega_x)$.

(Proof: $\rho_t x$ depends continuously on t for fixed x, according to 13.6.)

13.8 CONSTRUCTION of simultaneous approximations. Assume given a continuous $f: C[-1,0] \to R$. For every $x \in C[-1,0]$ and every $\lambda > 0$ we construct the simultaneous approximations $x_\lambda: R \to R$ as follows.

Let

13.8.1 $$x_\lambda(s) = \begin{cases} x(-1) & \text{for } s < -1 \\ x(s) & \text{for } -1 \leq s < 0; \end{cases}$$

then define

13.8.2 $$x_\lambda(s) = x(0) + \int_0^s f(\rho_{t-\lambda} x_\lambda)dt$$

first for $0 \leq s < \lambda$ (notice that then $\rho_{t-\lambda} x_\lambda(\theta) = x_\lambda(t-\lambda+\theta)$ has $t - \lambda + \theta \leq \lambda - \lambda + 0 = 0$); and using this, also for $\lambda \leq s < 2\lambda$, $2\lambda \leq s < 3\lambda$, etc.

We observe that each $x_\lambda: R \to R$ is well defined and continuous; and that

13.8.3 $$|x_\lambda(s_1) - x_\lambda(s_2)| \leq \int_{s_2}^{s_1} |f(\rho_{t-\lambda} x_\lambda)|dt$$

for $s_1 \geq s_2$ in R^+ (of course, for $s_1 \geq s_2$ in R^- the difference is 0).

It may be useful to note the similarities between 13.7.1 and 13.8.2: the former motivates the latter.

13.9 LEMMA. If $f: C[-1,0] \to R$ is continuous and bounded, then to every $x \in C[-1,0]$ there exists a solution x_0 of 13.2, with initial value x, and defined on R (i.e. with $\omega_{x_0} = +\infty$).

Proof. Set $\alpha = \sup|f|$, so that $\alpha < +\infty$ by assumption. Now perform the construction 13.8 (for the given x and all $\lambda > 0$). Then from 13.8.3,

$$|x_\lambda(s_1)-x_\lambda(s_2)| \leq \alpha|s_1-s_2|$$

for s_k in R^+, and actually even for s_k in R. Since α is independent of λ, the system $\{x_\lambda|\lambda > 0\}$ is equicontinuous; also, $x_\lambda(0) = x(0)$ is independent of λ. Hence, according to Ascoli's Theorem, there exists a sequence $\lambda_n \to 0$ such that $\{x_\lambda|\lambda = \lambda_n\}$ converges to some continuous $x_0: R \to R$ uniformly on compact sets. Then we apply 13.6 to $\rho_{t-\lambda_n}x_{\lambda_n}$ and obtain from 13.8.2 that

$$x_0(s) = x(0) + \int_0^s f(\rho_t x_0)dt \quad \text{for} \quad s \in R^+,$$

$$x_0(s) = x(s) \qquad\qquad \text{for} \quad s \in [-1,0]$$

(see 13.8.1 and continuity for the latter relation). Hence and from 13.7, $x_0: R \to R$ is a solution of 13.2 with initial value x. Q.E.D.

13.10 LEMMA. Assumptions: For $n = 1,2,\ldots$ and also $n = \infty$, g_n is a continuous map $C[-1,0] \to R$; $g_n \to g_\infty$ uniformly on bounded subsets; g_0 is bounded on bounded subsets.

Conclusion: To each $\xi \in R^+$ there exists $\omega > 0$ with the following property. Whenever $x_n \to x_\infty$ in $C[-1,0]$ with $\|x_n\| \leq \xi$, for any solutions $y_n: R \to R$ of 13.2 (with f replaced by g_n) with initial value x_n, there exists a subsequence y_m such that

$$y_m \to y_\infty \text{ uniformly on compact subsets of } (-\infty,\omega),$$

and that y_∞ is a solution of 13.2 (with f replaced by g_∞) with initial value x_∞.

Proof. This will follow from Ascoli's Theorem, essentially in the same manner as 13.9. However, we first need a common bound to $f_n(\rho_t y_n)$.

13.10.1 Set $\eta = \xi + 1$. We intend to show that there exists $\omega > 0$, such that, whenever x_n are as indicated in the assertion, then $\|\rho_t y_n\| \leq \eta$ for all $t \in [0,\omega]$; i.e.,

$$|y_n(s)| \leq \eta \quad \text{for all} \quad s \in [-1,\omega]$$

and all $n = 1,2,\dots$. Observe that this is so for $s \in [-1,0]$, since then

$$|y_n(s)| = |x_n(s)| \leq \xi < \xi + 1 = \eta.$$

Assume that this assertion does not hold. Then for small $\omega > 0$ there exists an index n and a point $s \in (0,\omega]$ (the dependence of n and s on ω is not explicitly indicated), such that $|y_n(s)| = \eta$. We may assume that $s \in R^+$ is the first point with this property, so that

13.10.2 $$\|\rho_t y_n\| \leq \eta \quad \text{for} \quad t \in [0,s].$$

Now,

$$|y_n(s)-y_n(0)| \geq |y_n(s)| - |x_n(0)| \geq \eta - \xi = 1.$$

Second, from the Mean Value Theorem, there exists $\sigma \in (0,s)$ with

$$y_n(s) - y_n(0) = s \cdot y_n'(\sigma) = s \cdot g_n(\rho_\sigma y_n).$$

Thus

$$1 \leq s \cdot |g_n(\rho_\sigma y_n)|.$$

Now observe that $\rho_\sigma y_n$ are uniformly bounded by 13.10.2; that then $g_n(\rho_\sigma(y_n))$ are bounded by assumption on g_n and g_∞; but that $s < \omega$ is arbitrarily small. This contradiction proves 13.10.1.

13.10.3 Set

$$\alpha = \sup\{g_n(y): \|y\| \leq \eta, \ n = 1,2,\ldots\};$$

by assumption on g_n and g_∞, $\alpha < +\infty$. In 13.10.1 we have shown that if y_n, y are as described in 13.10, then $\|\rho_t y_n\| \leq \eta$, and hence

$$|g_n(\rho_t y_n)| \leq \alpha \quad \text{for} \quad t \in [0,\omega]$$

and all $n = 1,2,\ldots$. Now apply 13.7 (with f replaced by g_n):

$$|y_n(s_1) - y_n(s_2)| \leq \alpha |s_1 - s_2|$$

for s_k in $[0,\omega]$. Thus the sequence y_n is equicontinuous in $[0,\omega]$, and we may proceed as in 13.9, via Ascoli's Theorem, to obtain our assertion (of course, in $[-1,0]$ we already have uniform convergence by assumption on the x_n and x_∞).

13.11 Proof of 13.5.1. First define, for $n = 1,2,\ldots,$ mappings f_n in the following fashion. By assumption f is bounded on the closed ball

$$S_n = \{x \in C[-1,0]: \|x\| \leq n\};$$

by Tietze's Theorem, there exists a continuous and bounded extension $f_n: C[-1,0] \to R$ of $f|S_n$. Then each f_n is bounded, and $f_n \to f$ uniformly on bounded subsets of $C[-1,0]$. According to Lemma 13.9, to arbitrarily given initial value $x \in C[-1,0]$ there exists a solution of 13.2 (with f replaced by f_n) defined on the entire R. Thence and from Lemma 13.10, to any $x \in C[-1,0]$ there exists $\varepsilon > 0$ (in 13.10 take $\xi = 1 + \|x\|$ and $\varepsilon = \min \omega, 1)$ such that $y\pi t$ is defined for all $t \in [0,\varepsilon)$ and all y with $\|y-x\| < \varepsilon$ (indeed, then $\|y\| \leq \varepsilon + \|x\| \leq \xi$).

13.12 Thus it only remains to prove continuity of π in the indicated domain. Thus, let $x_n \to x$ and $t_n \to t$. Consider any subsequence x_m, t_m. According to 13.10 there exists a sub-sequence x_k, t_k such that $x_k\pi t_k$ converges, necessarily to $x\pi t$ since this is the only solution through x. Since the limit is independent of the original subsequences chosen (i.e. x_m, t_m), we conclude that in the original sequence $x_n\pi t_n \to x\pi t$. Q.E.D.

13.13 Remarks.

13.13.1 It seems probable that in Theorem 13.4 one cannot simply omit the assumption that f be bounded on bounded subsets.

13.13.2 On the other hand, for those lsd-systems which are defined by functional-differential equations satisfying the hypotheses of Theorem 13.4, something more can be proved in connection with Lemma 1.10 (essentially this is a consequence of Lemma 13.10): If $\omega_x < +\infty$, then the net $\{x\pi t\colon 0 \le t < \omega_x\}$ is ultimately outside every weakly compact subset; in other words, $\|x\pi t\| \to +\infty$ as $t \to \omega_x$.

13.13.3 If $f\colon C[-1,0]$ is Lipschitzian, then π is global. In particular this applies to linear mappings f; however, this latter situation is a special case of [11, Theorem 5].

13.13.4 Let $f\colon C[-1,0]$ be locally Lipschitzian: to every $x \in C[-1,0]$ there exist $\varepsilon > 0 < \alpha$ such that

$$|f(x_1)-f(x_2)| \le \alpha\|x_1-x_2\|$$

whenever $\|x_k-x\| < \varepsilon$. Then f is continuous, and for

13.2 one can prove positive uniqueness; however, f need not be bounded on bounded subsets, so that one cannot apply Theorem 13.4 directly. Nevertheless we can prove 13.5.1 again by relativizing to the ε-ball centered at x, wherein f is globally Lipschitzian. Thus we have the following

13.14 THEOREM. Let $f: C[-1,0] \to R$ be locally Lipschitzian. Then the partial map π defined in 13.3.6 is an lsd-system on $C[-1,0]$; furthermore, each right solution ${}_x\pi$ of π is a solution of 13.2 with initial value x.

REFERENCES

[1] S. Ahmad, Dynamical systems of characteristic 0^+, to appear.

[2] J. Auslander, N. P. Bhatia and P. Seibert, Attractors in dynamical systems, Bol. Soc. Mat. Mexicana 9 (1964), 55-66.

[3] N. P. Bhatia, Weak attractors in dynamical systems , Bol. Soc. Mat. Mexicana 11 (1966), 56-64.

[4] ________, Semi-dynamical flow near a compact invariant set; Topological Dynamics (J. Auslander and W. H. Gottschalk, ed.), Proceedings of the Fort Collins 1967 symposium, Benjamin, New York, 1968.

[5] ________, Semi-dynamical systems, Lectures given at the International Summer School in Mathematical Systems Theory and Economics, June 1967, Varenna, Italy. Proceedings to appear in Lecture Notes in Operations Research and Mathematical Economics, Springer-Verlag, Berlin-Heidelberg-New York.

[6] ________, and G. P. Szegö, Dynamical Systems: Stability Theory and Applications, Lecture Notes in Mathematics 35, Springer-Verlag, Berlin-Heidelberg-New York, 1967.

[7] O. Hajek, Critical points of abstract dynamical systems, Comment. Math. Univ. Carol. 5 (1964), 121-124.

[8] ________, Structure of dynamical systems, ibid. 6 (1965), 53-72.

[9] ________, Local characterisation of local semi-dynamical systems, Math. Syst. Theory 2 (1968), 17-25.

[10] ________, Dynamical Systems in the Plane, Academic Press, New York-London, 1968.

[11] ________, Linear semi-dynamical systems, Mathematical Systems Theory, 2 (1968), 195-202.

[12] J. Hocking and G. S. Young, Topology, Addison-Wesley, Reading, 1961.

[13] L. Kelley, General Topology, Van Nostrand, Princeton, 1955.

[14] I. Kimura and T. Ura, Sur le courant extérieur à une région invariante; Théorème de Bendixson, Comment. Math. Univ. Santi Pauli 8 (1960), 23-39.

[15] V. V. Nemyckij and V. V. Stepanov, Qualitative Theory of Differential Equations, (translation), Princeton Mathematical Series, 22, Princeton University Press, Princeton, 1960.

[16] T. Ura, Sur le courant exterieur a une region invariante, Funkc. Ekvac. 2 (1959), 143-200.

[17] P. N. Bajaj, Some aspects of the semi-dynamical systems theory, Ph. D. Thesis, Case Western Reserve University, 1968.

INDEX

$A(M)$ 8.5.2
$A_s(M)$ 8.5.3
$A_u(M)$ 8.5.4
$A_w(M)$ 8.5.1
α_σ 2.13.2
α_x 2.13.1
attraction 8.4.2
—region of 8.5
—strong 8.4.3
—uniform 8.4.4
—weak 8.4.1
attractor 8.7

C_x 1.12
characteristic 0 12.11
cycle 1.15.2

D_x 6.1.1
dynamical system 0.2
—bilateral 0.2
—global 0.2
—local 0.6
—local semi 1.1, 0.7

E^n 0.13
escape time 1.1.1
—negative 2.13

$F_t(x)$ 2.14.3
$F_s^t(x)$ 2.14.2
$F(x)$ 2.14.1
funnel 2.14

group of periods 1.13

instability 9.2.1
—complete 9.2.3
—complete weak 9.2.2

invariance
—positive 3.1.1
—strong 3.1.3
—strong negative 3.1.2
—weak 3.7
—weak negative 3.7

J_x 6.1.2

K_x 5

L_σ^- 5.15
L_x 5.1
Liapunov function 10.1
limit set 5.1
lsd-system 1.1
—global 1.3.1

minimality
—positive 12.1
—strong 12.1
—weak 12.1

negative
—globality 1.3.2
—local existence 1.3.3
—unicity 1.3.4

ω_x 1.1.1
openness property 1.3.5

π 1.1
point
—critical 1.15.3
—periodic 1.15.2
—singular 1.6.2
—start 1.6.1
prolongation 6.1.1
prolongational limit 6.1.2

R 0.13

R^+ 0.13

R^- 0.13

R^n 0.13

S^n 0.13

solution 2.1
—left 2.5.1
—left maximal 2.5.3
—principal 2.7
—right 1.12.2, 2.5.2
—right maximal 2.5.4

stability 7.1.1
—asymptotic 8.16
—orbital 7.1.2

trajectory
—complete 2.12
—negative 2.7
—positive 1.12
—principal 2.7, 2.12
—self-intersecting 1.15.1

x^π 1.12.2

Offsetdruck: Julius Beltz, Weinheim/Bergstr.

Lecture Notes in Mathematics

Bisher erschienen/Already published

Vol. 1: J. Wermer, Seminar über Funktionen-Algebren. IV, 30 Seiten. 1964. DM 3,80 / 0.95

Vol. 2: A. Borel, Cohomologie des espaces localement compacts d'après J. Leray. IV, 93 pages. 1964. DM 9,– / $ 2.25

Vol. 3: J. F. Adams, Stable Homotopy Theory. 2nd. revised edition. IV, 78 pages. 1966. DM 7,80 / $ 1.95

Vol. 4: M. Arkowitz and C. R. Curjel, Groups of Homotopy Classes. 2nd. revised edition. IV, 36 pages. 1967. DM 4,80 / $ 1.20

Vol. 5: J.-P. Serre, Cohomologie Galoisienne. Troisième édition. VIII, 214 pages. 1965. DM 18,– / $ 4.50

Vol. 6: H. Hermes, Eine Termlogik mit Auswahloperator. IV, 42 Seiten. 1965. DM 5,80 / $ 1.45

Vol. 7: Ph. Tondeur, Introduction to Lie Groups and Transformation Groups. VIII, 176 pages. 1965. DM 13,50 / $ 3.40

Vol. 8: G. Fichera, Linear Elliptic Differential Systems and Eigenvalue Problems. IV, 176 pages. 1965. DM 13,50 / $ 3.40

Vol. 9: P. L. Ivănescu, Pseudo-Boolean Programming and Applications. IV, 50 pages. 1965. DM 4,80 / $ 1.20

Vol. 10: H. Lüneburg, Die Suzukigruppen und ihre Geometrien. VI, 111 Seiten. 1965. DM 8,– / $ 2.00

Vol. 11: J.-P. Serre, Algèbre Locale. Multiplicités. Rédigé par P. Gabriel. Seconde édition. VIII, 192 pages. 1965. DM 12,– / $ 3.00

Vol. 12: A. Dold, Halbexakte Homotopiefunktoren. II, 157 Seiten. 1966. DM 12,– / $ 3.00

Vol. 13: E. Thomas, Seminar on Fiber Spaces. IV, 45 pages. 1966. DM 4,80 / $ 1.20

Vol. 14: H. Werner, Vorlesung über Approximationstheorie. IV, 184 Seiten und 12 Seiten Anhang. 1966. DM 14,– / $ 3.50

Vol. 15: F. Oort, Commutative Group Schemes. VI, 133 pages. 1966. DM 9,80 / $ 2.45

Vol. 16: J. Pfanzagl and W. Pierlo, Compact Systems of Sets. IV, 48 pages. 1966. DM 5,80 / $ 1.45

Vol. 17: C. Müller, Spherical Harmonics. IV, 46 pages. 1966. DM 5,– / $ 1.25

Vol 18: H.-B. Brinkmann und D. Puppe, Kategorien und Funktoren. XII, 107 Seiten, 1966. DM 8,– / $ 2.00

Vol. 19: G. Stolzenberg, Volumes, Limits and Extensions of Analytic Varieties. IV, 45 pages. 1966. DM 5,40 / $ 1.35

Vol. 20: R. Hartshorne, Residues and Duality. VIII, 423 pages. 1966. DM 20,– / $ 5.00

Vol. 21: Seminar on Complex Multiplication. By A. Borel, S. Chowla, C. S. Herz, K. Iwasawa, J.-P. Serre. IV, 102 pages. 1966. DM 8,– / $ 2.00

Vol. 22: H. Bauer, Harmonische Räume und ihre Potentialtheorie. IV, 175 Seiten. 1966. DM 14,– / $ 3.50

Vol. 23: P. L. Ivănescu and S. Rudeanu, Pseudo-Boolean Methods for Bivalent Programming. 120 pages. 1966. DM 10,– / $ 2.50

Vol. 24: J. Lambek, Completions of Categories. IV, 69 pages. 1966. DM 6,80 / $ 1.70

Vol. 25: R. Narasimhan, Introduction to the Theory of Analytic Spaces. IV, 143 pages. 1966. DM 10,– / $ 2.50

Vol. 26: P.-A. Meyer, Processus de Markov. IV, 190 pages. 1967. DM 15,– / $ 3.75

Vol. 27: H. P. Künzi und S. T. Tan, Lineare Optimierung großer Systeme. VI, 121 Seiten. 1966. DM 12,– / $ 3.00

Vol. 28: P. E. Conner and E. E. Floyd, The Relation of Cobordism to K-Theories. VIII, 112 pages. 1966. DM 9,80 / $ 2.45

Vol. 29: K. Chandrasekharan, Einführung in die Analytische Zahlentheorie. VI, 199 Seiten. 1966. DM 16,80 / $ 4.20

Vol. 30: A. Frölicher and W. Bucher, Calculus in Vector Spaces without Norm. X, 146 pages. 1966. DM 12,– / $ 3.00

Vol. 31: Symposium on Probability Methods in Analysis. Chairman. D. A. Kappos. IV, 329 pages. 1967. DM 20,– / $ 5.00

Vol. 32: M. André, Méthode Simpliciale en Algèbre Homologique et Algèbre Commutative. IV, 122 pages. 1967. DM 12,– / $ 3.00

Vol. 33: G. I. Targonski, Seminar on Functional Operators and Equations. IV, 110 pages. 1967. DM 10,– / $ 2.50

Vol. 34: G. E. Bredon, Equivariant Cohomology Theories. VI, 64 pages. 1967. DM 6,80 / $ 1.70

Vol. 35: N. P. Bhatia and G. P. Szegö, Dynamical Systems. Stability Theory and Applications. VI, 416 pages. 1967. DM 24,– / $ 6.00

Vol. 36: A. Borel, Topics in the Homology Theory of Fibre Bundles. VI, 95 pages. 1967. DM 9,– / $ 2.25

Vol. 37: R. B. Jensen, Modelle der Mengenlehre. X, 176 Seiten. 1967. DM 14,– / $ 3.50

Vol. 38: R. Berger, R. Kiehl, E. Kunz und H.-J. Nastold, Differentialrechnung in der analytischen Geometrie IV, 134 Seiten. 1967. DM 12,– / $ 3.00

Vol. 39: Séminaire de Probabilités I. II, 189 pages. 1967. DM 14,– / $ 3.50

Vol. 40: J. Tits, Tabellen zu den einfachen Lie Gruppen und ihren Darstellungen. VI, 53 Seiten. 1967. DM 6.80 / $ 1.70

Vol. 41: A. Grothendieck, Local Cohomology. VI, 106 pages. 1967. DM 10.– / $ 2.50

Vol. 42: J. F. Berglund and K. H. Hofmann, Compact Semitopological Semigroups and Weakly Almost Periodic Functions. VI, 160 pages. 1967. DM 12,– / $ 3.00

Vol. 43: D. G. Quillen, Homotopical Algebra VI, 157 pages. 1967. DM 14,– / $ 3.50

Vol. 44: K. Urbanik, Lectures on Prediction Theory IV, 50 pages. 1967. DM 5,80 / $ 1.45

Vol. 45: A. Wilansky, Topics in Functional Analysis VI, 102 pages. 1967. DM 9,60 / $ 2.40

Vol. 46: P. E. Conner, Seminar on Periodic Maps IV, 116 pages. 1967. DM 10,60 / $ 2.65

Vol. 47: Reports of the Midwest Category Seminar I. IV, 181 pages. 1967. DM 14,80 / $ 3.70

Vol. 48: G. de Rham, S. Maumary et M. A. Kervaire, Torsion et Type Simple d'Homotopie. IV, 101 pages. 1967. DM 9,60 / $ 2.40

Vol. 49: C. Faith, Lectures on Injective Modules and Quotient Rings. XVI, 140 pages. 1967. DM 12,80 / $ 3.20

Vol. 50: L. Zalcman, Analytic Capacity and Rational Approximation, VI, 155 pages. 1968. DM 13.20 / $ 3.40

Vol. 51: Séminaire de Probabilités II. IV, 199 pages. 1968. DM 14,– / $ 3.50

Vol. 52: D. J. Simms, Lie Groups and Quantum Mechanics. IV, 90 pages. 1968. DM 8,– / $ 2.00

Vol. 53: J. Cerf, Sur les difféomorphismes de la sphère de dimension trois ($\Gamma_4 = O$). XII, 133 pages. 1968. DM 12,– / $ 3.00

Vol. 54: G. Shimura, Automorphic Functions and Number Theory VI, 69 pages. 1968. DM 8,– / $ 2.00

Vol. 55: D. Gromoll, W. Klingenberg und W. Meyer Riemannsche Geometrie im Großen VI, 287 Seiten. 1968. DM 20,– / $ 5.00

Bitte wenden / Continued